METHODS IN MOLECULAR BIOLOGY™

Series Editor
John M. Walker
School of Life Sciences
University of Hertfordshire
Hatfield, Hertfordshire, AL10 9AB, UK

For further volumes:
http://www.springer.com/series/7651

Imaging and Tracking Stem Cells

Methods and Protocols

Edited by

Kursad Turksen

Ottawa Hospital Research Institute
Sprott Centre for Stem Cell Research
Regenerative Medicine Program
Smyth Road 501
K1Y 8L6 Ottawa, Ontario
Canada

Humana Press

Editor
Kursad Turksen
Ottawa Hospital Research Institute
Sprott Centre for Stem Cell Research
Regenerative Medicine Program
Smyth Road 501
K1Y 8L6 Ottawa, Ontario
Canada

ISBN 978-1-4939-6059-0   ISBN 978-1-62703-559-0 (eBook)
DOI 10.1007/978-1-62703-559-0
Springer New York Heidelberg Dordrecht London

Preface

Amongst the most crucial advances contributing to our increased understanding of stem cells have been those related to imaging and lineage tracing, including live cells, in vivo and in vitro. I have gathered here a select series of representative protocols that hopefully will provide both a flavor of the field as it currently is and hopefully stimulate new approaches and methodologies by those interested in tracking stem cells and their progeny.

The protocols in this volume are faithful to the mission statement of the Methods in Molecular Biology series: They have been validated for reproducibility and are described in an easy-to-follow, step-by-step fashion so as to be valuable for not only experts but also novices in the stem cell field. That goal is achieved because of the generosity of the contributors who have carefully described their protocols in this volume, and I thank them for their efforts.

My thanks as well go to Dr. John Walker, the Editor in Chief of the Methods in Molecular Biology series, for his support during the course of preparing this volume.

I am also grateful to Patrick Marton, the Editor of Methods in Molecular Biology and Springer Protocols series, for his continuous support from idea to completion of this volume.

Finally, I would like to thank David Casey for his outstanding help during the production of this volume.

Ottawa, ON, Canada   *Kursad Turksen*

Contents

Contributors

TAKAYA ABE • *Laboratory for Animal Resources and Genetic Engineering, RIKEN Center for Developmental Biology (CDB), Kobe, Japan*

JOANNA ADAMCZAK • *In-vivo-NMR Laboratory, Max Planck Institute for Neurological Research, Cologne, Germany*

SHINICHI AIZAWA • *Laboratory for Animal Resources and Genetic Engineering, RIKEN Center for Developmental Biology (CDB), Kobe, Japan*

MARIE-THERESE ARMENTERO • *Laboratory of Functional Neurochemistry, C. Mondino National Institute of Neurology Foundation, IRCCS, Pavia, Italy*

MARKUS ASWENDT • *In-vivo-NMR Laboratory, Max Planck Institute for Neurological Research, Cologne, Germany*

MARZIA BELICCHI • *Stem Cell Laboratory, Dipartimento di Fisiopatologia Medico-Chirurgica e Dei Trapianti, Fondazione IRCCS Ca' Granda Ospedale Maggiore Policlinico, Università degli Studi di Milano, Milano, Italy*

ZWI BERNEMAN • *Laboratory of Experimental Hematology, Vaccine and Infectious Disease Institute (vaxinfectio), University of Antwerp, Antwerp, Belgium*

BENEDIKT BERNINGER • *Department of Physiological Genomics, Institute of Physiology, Ludwig-Maximilians University Munich, Munich, Germany*

SUZANNE E. BERRY • *Department of Comparative Biosciences, University of Illinois at Urbana-Champaign, Urbana, IL, USA*

MARNI D. BOPPART • *Department of Kinesiology and Community Health, University of Illinois at Urbana-Champaign, Urbana, IL, USA*

STEPHEN A. BOPPART • *Biophotonics Imaging Laboratory, Department of Electrical and Computer Engineering, Department of Bioengineering, & Department of Internal Medicine, Beckman Institute for Advanced Science and Technology, University of Illinois at Urbana-Champaign, Urbana, IL, USA*

PATRIZIA BOSSOLASCO • *Dipartimento di Farmacologia, Chemioterapia e Tossicologia Medica, Fondazione Matarelli, Università degli Studi di Milano, Milan, Italy*

ANDREW J. BOWER • *Biophotonics Imaging Laboratory, Department of Electrical and Computer Engineering, Beckman Institute for Advanced Science and Technology, University of Illinois at Urbana-Champaign, Urbana, IL, USA*

ROSTISLAV CASTILLO • *Department of Radiology, Molecular Imaging Program at Stanford, Stanford University, Stanford, CA, USA*

JU LAN CHUN • *Department of Animal Sciences, University of Illinois at Urbana-Champaign, Urbana, IL, USA*

MARCOS R. COSTA • *Brain Institute, Federal University of Rio Grande do Norte, Natal, Brazil*

LIDIA COVA • *Department of Neurology and Laboratory of Neuroscience, IRCCS Istituto Auxologico Italiano, Cusano Milanino, Italy*

JASMIJN DAANS • *Laboratory of Experimental Hematology, Vaccine and Infectious Disease Institute (vaxinfectio), University of Antwerp, Antwerp, Belgium*

HEIKE E. DALDRUP-LINK • *Department of Radiology, Molecular Imaging Program at Stanford, Stanford University, Stanford, CA, USA*

NATHALIE DE VOCHT • *Bio-Imaging Laboratory, University of Antwerp, Antwerp, Belgium*

ANDREA FARINI • *Stem Cell Laboratory, Dipartimento di Fisiopatologia Medico-Chirurgica e Dei Trapianti, Fondazione IRCCS Ca' Granda Ospedale Maggiore Policlinico, Università degli Studi di Milano, Milano, Italy*

GÁBOR FÖLDES • *Imperial Centre for Experimental and Translational Medicine, National Heart and Lung Institute, Imperial College London, London, UK*

TOSHIHIKO FUJIMORI • *Division of Embryology, National Institute for Basic Biology (NIBB), Okazaki, Japan*

BENEDIKT W. GRAF • *Biophotonics Imaging Laboratory, Department of Electrical and Computer Engineering, Beckman Institute for Advanced Science and Technology, University of Illinois at Urbana-Champaign, Urbana, IL, USA*

ANNA-KATERINA HADJANTONAKIS • *Developmental Biology Program, Sloan-Kettering Institute, New York, NY, USA*

ZHONGCHAO HAN • *State Key Laboratory of Hematology, Chinese Academy of Medical Sciences, Tianjin, China*

ZUOXIANG HE • *Department of Cardiac Nuclear Imaging, Peking Union Medical College and Chinese Academy of Medical Sciences, Fuwai Hospital, Beijing, China*

MATHIAS HOEHN • *In-vivo-NMR Laboratory, Max Planck Institute for Neurological Research, Cologne, Germany*

CHLOÉ HOORNAERT • *Laboratory of Experimental Hematology, Vaccine and Infectious Disease Institute (vaxinfectio), University of Antwerp, Antwerp, Belgium*

MINJUNG KANG • *Developmental Biology Program, Sloan-Kettering Institute, New York NY, USA*

JOSEPH J. KIM • *Department of Biomedical Engineering, Chemical and Biochemical Engineering, Rutgers University, Piscataway, NJ, USA*

DEBBIE LE BLON • *Laboratory of Experimental Hematology, Vaccine and Infectious Disease Institute (vaxinfectio), University of Antwerp, Antwerp, Belgium*

LIANG LENG • *Department of Pathophysiology, Nankai University School of Medicine, Tianjin, China*

ZONGJIN LI • *Department of Pathophysiology, Nankai University, Tianjin, China*

XINGHUA LOU • *Developmental Biology Program, Sloan-Kettering Institute, New York, NY, USA*

MIRELLA MEREGALLI • *Stem Cell Laboratory, Dipartimento di Fisiopatologia Medico-Chirurgica e Dei Trapianti, Fondazione IRCCS Ca' Granda Ospedale Maggiore Policlinico, Università degli Studi di Milano, Milano, Italy*

MAXIME MIOULANE • *Imperial Centre for Experimental and Translational Medicine, National Heart and Lung Institute, Imperial College London, London, UK*

PRABHAS V. MOGHE • *Department of Biomedical Engineering, Chemical and Biochemical Engineering, Rutgers University, Piscataway, NJ, USA*

SILVIA MUÑOZ-DESCALZO • *Developmental Biology Program, Sloan-Kettering Institute, New York, NY, USA*

GO NAGAMATSU • *The Sakaguchi Laboratory, Department of Cell Differentiation, School of Medicine, Keio University, Tokyo, Japan; Precursory Research for Embryonic Science and Technology, Japan Science and Technology Agency, Saitama, Japan*

HOSSEIN NEJADNIK • *Department of Radiology, Molecular Imaging Program at Stanford, Stanford University, Stanford, CA, USA*

BORIS ODINTSOV • *Department of Bioengineering, Biomedical Imaging Center, Beckman Institute, University of Illinois at Urbana-Champaign, Urbana, IL, USA*

FELIPE ORTEGA • *Department of Physiological Genomics, Institute of Physiology, Ludwig-Maximilians University Munich, Munich, Germany*

PETER PONSAERTS • *Laboratory of Experimental Hematology, Vaccine and Infectious Disease Institute (vaxinfectio), University of Antwerp, Antwerp, Belgium*

JELLE PRAET • *Laboratory of Experimental Hematology, Vaccine and Infectious Disease Institute (vaxinfectio), University of Antwerp, Antwerp, Belgium*

KRISTIEN REEKMANS • *Laboratory of Experimental Hematology, Vaccine and Infectious Disease Institute (vaxinfectio), University of Antwerp, Antwerp, Belgium*

NORA SANDU • *Department of Neurosurgery, University of Paris, Paris, France*

BERNHARD SCHALLER • *Department of Neurosurgery, University of Paris, Paris, France; Department of Neurosurgery, Tokuda Hospital Sofia, Sofia, Bulgaria*

TOMA SPIRIEV • *Department of Neurosurgery, Tokuda Hospital Sofia, Sofia, Bulgaria*

WEIJUN SU • *Department of Pathology, Nankai University School of Medicine, Tianjin, China*

TOSHIO SUDA • *The Sakaguchi Laboratory, Department of Cell Differentiation, School of Medicine, Keio University, Tokyo, Japan*

JIAN TAJBAKHSH • *Translational Cytomics Group and Chromatin Biology Laboratory, Department of Surgery, Cedars-Sinai Medical Center, Los Angeles, CA, USA*

ANNETTE TENNSTAEDT • *In-vivo-NMR Laboratory, Max Planck Institute for Neurological Research, Cologne, Germany*

YVAN TORRENTE • *Stem Cell Laboratory, Dipartimento di Fisiopatologia Medico-Chirurgica e Dei Trapianti, Fondazione IRCCS Ca' Granda Ospedale Maggiore Policlinico, Università degli Studi di Milano, Milano, Italy*

ANNEMIE VAN DER LINDEN • *VochtBio-Imaging Laboratory, University of Antwerp, Antwerp, Belgium*

SEBASTIÁN L. VEGA • *Chemical and Biochemical Engineering, Rutgers University, Piscataway, NJ, USA*

CHIARA VILLA • *Stem Cell Laboratory, Dipartimento di Fisiopatologia Medico-Chirurgica e Dei Trapianti, Fondazione IRCCS Ca' Granda Ospedale Maggiore Policlinico, Università degli Studi di Milano, Milano, Italy*

PANAGIOTIS XENOPOULOS • *Developmental Biology Program, Sloan-Kettering Institute, New York, NY, USA*

HUALEI ZHANG • *Laboratories of Molecular Imaging, Department of Radiology, School of Medicine, University of Pennsylvania, Philadelphia, PA, USA*

YOUBO ZHAO • *Biophotonics Imaging Laboratory, Beckman Institute for Advanced Science and Technology, University of Illinois at Urbana-Champaign, Urbana, IL, USA*

RONG ZHOU • *Laboratories of Molecular Imaging, Department of Radiology, School of Medicine, University of Pennsylvania, Philadelphia, PA, USA*

Methods Molecular Biology (2013) 1052: 1–11
DOI 10.1007/7651_2013_22
© Springer Science+Business Media New York 2013
Published online: 3 May 2013

Primary Culture and Live Imaging of Adult Neural Stem Cells and Their Progeny

Felipe Ortega, Benedikt Berninger, and Marcos R. Costa

Abstract

Adult neural stem cells (NSC) generate neurons throughout life, but little is known about the sequence of events involved in the transition from NSC to neurons. Studying the intermediary steps involved in the specification of neuronal cells from NSCs requires observation of cells in real time. Here we describe a primary culture of the adult subependymal zone (SEZ) which allows for continuous live imaging to characterize the mode of cell division and lineage progression of adult NSCs and their progeny. To this end, cells are cultured at low density under adherent conditions and without growth factors. Under these conditions, NSCs display classical hallmarks of adult SEZ NSCs in vivo, such as astroglial marker expression and promoter activity, a slow cell cycle, and a predominantly neurogenic potential. Video time-lapse microscopy experiments using this cell preparation allow for studying the steps involved in the generation of fast-dividing precursors and neuroblasts from slow-dividing astroglia/NSCs.

Keywords: Adult neural stem cell, Subependymal zone, Neurogenesis, Cell culture, Time-lapse video microscopy, Astroglia

1 Introduction

The adult olfactory bulb (OB) is constantly supplied with newly generated neurons (1) that are thought to derive from adult NSCs residing in distinct compartments of the wall of the lateral ventricle (adult subependymal zone, SEZ) and the rostral migratory stream (RMS) (2, 3). These neurons comprise distinct types of interneurons populating different layers within the OB, most of which use the transmitter GABA, while a small subpopulation belongs to the glutamatergic lineage (4–8). Moreover, besides these neurons also some oligodendroglial progeny is generated, but it is currently unclear whether these arise from the same or distinct population of NSCs. Thus, while there is evidence that these diverse progenies arise from cells with stem cell properties (characterized by their ability of self-renewal and by multipotency in the neurosphere assay in vitro) (9), little is known about the actual behavior of adult NSCs due to the fact that imaging these cells in vivo is so far an unmet challenge. Recently, clonal analysis study has been performed in the stem cell niche of the other neurogenic zone, the adult dentate gyrus, demonstrating the ability of individual radial glia-like neural stem cells (NSC) for both self-renewal and multi-lineage differentiation. But such clonal analysis approach requires that NSCs and their progeny exhibit only very restricted migratory behavior. Given the extensive migration of neuroblasts in the SEZ and the RMS, clonal analysis is virtually impossible in these neurogenic zones.

To study the behavior of adult NSCs and their progeny in vitro, research has largely relied on the neurosphere assay in which stem and progenitor cells are expanded in the presence of epidermal growth factor (EGF) and fibroblast growth factor-2 (FGF2) under non-adherent conditions (10, 11). This assay however presents several caveats (12). Firstly, neurosphere formation is initiated by not only stem but also transit-amplifying cells (13, 14). Secondly, stem cells in vivo divide slowly, while neurosphere cells are fast-dividing. Thirdly, neurosphere cells give rise predominantly to astroglia and only to a low degree to neurons, in sharp contrast to adult NSC in vivo (13, 15). Increased astrogliogenesis can be attributed to the treatment with EGF and FGF2 (13, 16), indicating that the very prerequisite of neurosphere formation results in aberrant stem cell behavior. Fourthly, due to the 3-dimensional compact structure of neurospheres the mode of cell division and lineage progression is difficult to image on a single-cell level.

To overcome some of the limitations of the neurosphere assay several laboratories developed adherent SEZ cultures that replicate some aspects of the SEZ in vivo (17, 18). Lim and Alvarez-Buylla developed an interesting strategy based on coculturing adult SEZ cells with postnatal astroglia, aiming thereby to mimic the SEZ stem cell niche comprising niche astrocytes (17). However, the cocultured astroglia secrete unidentified growth factors and extracellular matrix molecules and introduce direct contact-dependent cell-to-cell signaling obviously affecting the mode of cell division and lineage progression of adult NSCs.

In order to reduce these unknowns and potentially unphysiological side effects of cocultured cells and growth factor addition to a minimum, we developed a low-density primary culture of the adult SEZ which (a) permits single-cell tracking for studying the mode of cell division and lineage progression in the presence of a minimum of extrinsic stimuli, (b) allows for the direct observation of the alleged sequence of transitions from slow-cycling NSCs to transit-amplifying precursors and neuroblasts as in vivo, and (c) allows for studying the functional properties and cellular plasticity of their neuronal progeny (19). Here, we describe a complete protocol for how to isolate and culture adult SEZ cells in serum- and growth

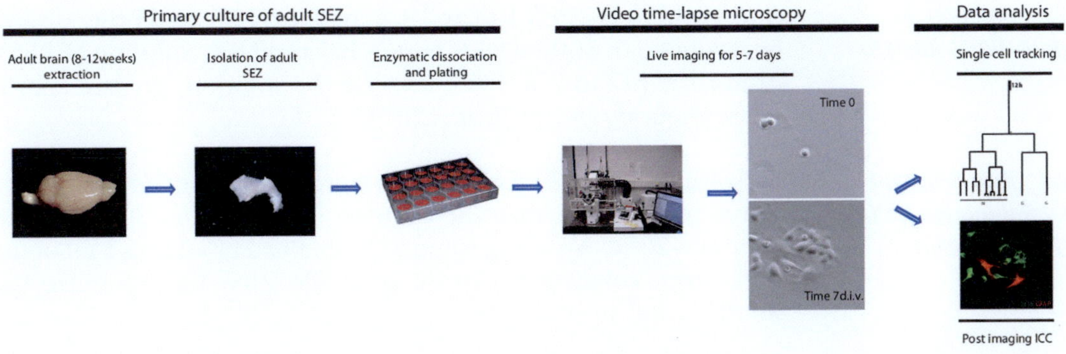

Fig. 1 Scheme demonstrating the major steps of the protocol. Displayed are the steps for isolation, primary culture, and live imaging of adult subependymal zone neural stem cells

factor-free medium and how to employ this preparation to perform time-lapse video microscopy experiments for subsequent tracking of individual lineages derived from neural stem and progenitor cells (Fig. 1).

2 Materials

2.1 Dissection Medium

Add 5 ml of HEPES buffer solution 1 M (final concentration: 10 mM) to 495 ml of Hanks' Balanced Salt Solution (HBSS) 1×. Store the dissection medium at 4 °C for up to 2 weeks.

2.2 Dissociation Solution

Dissolve 3.4 mg of trypsin powder and 3.5 mg of hyaluronidase powder in 5 ml of Solution I (see below). This solution must be prepared shortly before use.

2.3 B27-Supplemented Culture Medium

Add 1 ml of B27 serum-free supplement, 0.5 ml penicillin/streptomycin, 0.4 ml HEPES buffer solution 1 M, and GLUTAMAX at a final concentration of 2 mM and bring up to a final volume of 50 ml with Dulbecco's modified Eagle's medium—DMEM/F12 (1:1) medium.

2.4 Solution 1 (HBSS–Glucose)

Mix 50 ml of HBSS 10×, with 6 ml of 45 % D-(+)-glucose, and 7.5 ml of HEPES 1 M, and bring up to a final volume of 500 ml with pure H_2O. Adjust the pH to 7.5. Filter-sterilize the solution, prepare appropriate aliquots, and store at −20 °C.

2.5 Solution 2 (Saccharose–HBSS)

Add 25 ml of HBSS 10× and D-(+)-saccharose at a final concentration of 0.9 M (154 g), and bring up to a final volume of 500 ml with pure H_2O. Adjust the pH to 7.5. Filter-sterilize the solution, prepare appropriate aliquots, and store at −20 °C.

2.6 Solution 3 (BSA–EBSS–HEPES)

Add 10 ml of HEPES 1 M and 4 % (wt/vol) BSA, and bring up to a final volume of 500 ml with Earle's Balanced Salt Solution (EBSS). Adjust the pH to 7.5. Filter-sterilize the solution, prepare appropriate aliquots, and store at −20 °C.

2.7 Poly-ᴅ-Lysine Hydrobromide (PDL) Stock Solution

Dissolve 50 mg of PDL powder in pure H_2O at a concentration of 1 mg/ml, filter-sterilize the solution, prepare 1 ml aliquots, and store at −20 °C. Working solution is obtained by adding 1 ml of the PDL stock solution to 50 ml of sterile PBS. Filter-sterilize the solution, and store at 4 °C.

2.8 Poly-ᴅ-Lysine Hydrobromide (PDL)-Coating of Plastic Coverslips

Add 500 µl of PDL working solution to each well of a 24-well tissue culture plate containing plastic coverslips. Incubate the plate for at least 2 h or overnight at 37 °C. Wash thoroughly 3–4 times with pure water, and let the coverslips dry in a laminar airflow. Store coated plastic coverslips at 4 °C for no longer than 1 week.

2.9 Paraformaldehyde

To prepare 100 ml of paraformaldehyde 4 % (PFA 4 %) dilute 4 g of PFA in 40 ml of dH_2O. Warm up the solution at 70–80 °C using a hot plate and stir using a magnetic stirrer until the solution turns clear. Allow the solution to cool at room temperature under continued stirring. Add 50 ml of PBS 0.2 M and complete the volume to 100 ml with dH_2O. Adjust the pH to 7.4 and aliquot PFA 4 % solution in 15 ml conical tubes. Store at 4 °C up to 1 month.

2.10 Post-imaging Immunocytochemistry

After live image acquisition, fix cultures with 4 % PFA for 10 min and wash three times with 10 mM PBS. Primary antibodies can be incubated overnight at 4 °C or 2 h at room temperature and are detected by subclass-specific secondary antibodies conjugated with different fluorophores. Tissue plates are then repositioned in the microscope and to acquire fluorescent images that match the last live images taken before fixation.

2.11 Microscope

Inverted microscope with phase contrast and epifluorescence equipped with Long Distance Plan-Neofluar objectives (10× and 20×), an incubation system, scanning stage, and software for image acquisition and time-lapse experiments. We use a microscope Axio Observer (Zeiss) equipped with Incubator XL S1 and Heating Unit XL S (Zeiss); TempModule S1 (Zeiss); CO_2-Module S1 (Zeiss); Heating Insert P S and CO_2-Cover (Pecon); Scanning stage 130 × 85 (Zeiss); and Software AxioVision v4.8 (Zeiss).

2.12 Tracking System

In this protocol, we describe a method for analysis using the software Timm's Tracking Tool (TTT). Information about installation and use of TTT is available at http://www.helmholtz-muenchen. de/isf/haematopoese/software-download/index.html. However, tracking can be performed using other image editors, such as ImageJ (http://rsbweb.nih.gov/ij/).

2.13 Movie Editor Free version of ImageJ can be downloaded at http://rsbweb.nih. gov/ij/.

3 Methods

3.1 Primary Cultures of Neural Stem Cells Isolated from the Adult SEZ

1. During design and performance of any experiments that involve animal handling, governmental and institutional regulations regarding the use of animal for research purposes must be strictly followed.

2. Extract the brains of 8–12-week-old mice after decapitation and place them in a Petri dish (60 mmØ) filled with ice-cold dissection medium. The total number of animals required for a specific experiment should be calculated taking into account that one complete brain yields approximately $3–4 \times 10^4$ live cells after dissociation.

3. Using sterile material, cut the brain coronally at the level of the optic chiasm and subsequently separate both hemispheres using a surgical blade (Fig. 2). Uncover the ventricular face of the lateral wall by pulling from the caudal side with forceps as depicted in Fig. 2. Then with extreme care, in order not to damage the surface of the SEZ, cut the first around the perimeter of the SEZ and then just beneath the surface to obtain a thin sheath of tissue containing the lateral wall of the lateral ventricle. Try to avoid the attachment of adjacent striatal tissue. Collect the sheaths of lateral ventricular wall tissue in a 15 ml conical tube previously filled with 10 ml of dissection medium (Note 1).

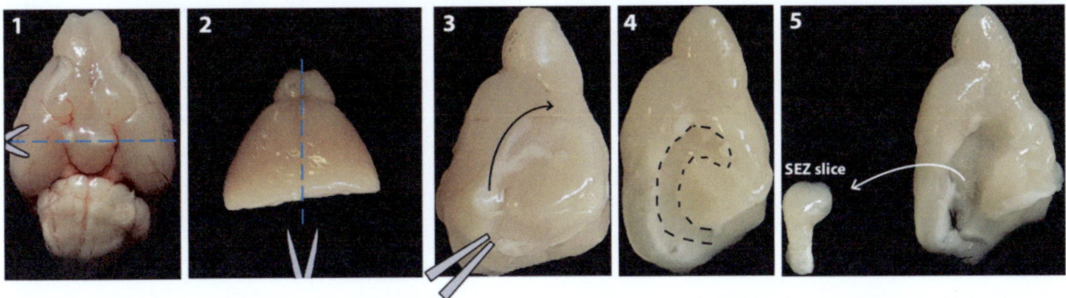

Fig. 2 Isolation of the adult SEZ. (*1*) Ventral view of an adult mouse brain after extraction showing the level for coronal transection (*dashed line*). (*2*) Dorsal view of the anterior part of the brain indicating the level for sagittal transection (*dashed line*). (*3*) Medial view of the *left* hemisphere. Note the closed ventricle. The ventricle should be opened from caudal to rostral using a forceps (depicted in *gray*); *arrow* indicates the direction to pull the medially located structures with the forceps, uncovering the lateral wall of the lateral ventricle. (*4*) View of the lateral wall of the lateral ventricle. The *outlined area* should be cut using an extra-fine scissor. (*5*) View of the isolated SEZ slice

4. Aspirate and discard the dissection medium with a sterile Pasteur pipette. Filter, using a 0.2 μm pore size syringe filters and a syringe, 5 ml of dissociation medium previously warmed at 37 °C and add it to the conical tube. Incubate the solution for 15 min at 37 °C. Next, triturate the tissue carefully up and down using a 5 ml pipette. The aim of this step is to release into the solution the NSC located just beneath the surface of the ventricle wall; therefore a complete dissociation is not required and the trituration should be stopped as soon as the medium becomes turbid. In any case no more than ten times of single up and down trituration steps are necessary. Then, incubate the solution for another 15 min at 37 °C (Notes 2a and 2b).

5. The dissociation medium should be inactivated after the incubation by the addition of 5 ml of ice-cold solution 3. Homogenize the solution by mixing up and down ten times with a 10 ml disposable pipette. Subsequently, filter the suspension using a 70 μm cell strainer in order to eliminate the tissue pieces that were not dissociated during step 4. Transfer the filtered solution to a new 15 ml conical tube.

6. Centrifuge at 200 × g for 5 min at 4 °C.

7. Aspirate and discard the supernatant, and then resuspend the cells in 10 ml of ice-cold solution 2.

8. Centrifuge at 450 × g for 10 min at 4 °C.

9. Aspirate and discard the supernatant, then resuspend the cells in 2 ml of ice-cold solution 3, and transfer it gently on top of a 15 ml conical tube previously filled with 12 ml of ice-cold solution 3.

10. Centrifuge at 250 × g for 7 min at 4 °C.

11. Aspirate and discard the supernatant, and then resuspend the cells in B27-supplemented culture medium (previously warmed at 37 °C) (Note 2c).

12. Count live cells using a Neubauer chamber and trypan blue to label dead cells. Seed cells at density of 200–300 cells per mm^2 (usually this represents an entire brain per well) in a 24-well tissue culture plate previously coated with poly-D-lysine. Leave the plate inside the incubator for 2 h to allow for proper attachment of the cells (Notes 2d and 2e).

3.2 Performing Video Time-Lapse Microscopy

1. Before starting a video time-lapse experiment, make sure that CO_2 (8 %) and temperature (37 °C) conditions are stable, in order to guarantee the viability of the culture. Therefore, the CO_2 and temperature modules, the computer, the microscope, and fluorescence lamp (the latter only when fluorescence is required) devices should be switched on in advance. Specific software, in our case AxioVision, will control incubation conditions (Note 2f).

2. Once the temperature and CO_2 conditions are stable the time-lapse experiment can be performed. Prior to placing the culture plate on the microscope stage, a mark should be made with a pen on the bottom of the plate. This mark will be used to set the xyz zero-position in the AxioVision software (Microscope Menu → Microscope → XYZ → Manlinkto set coordinates manually to zero). The zero position can be then easily found and used to reset the coordinates whenever necessary.

3. Next run the AxioVision application named "smart experiment" and set the general conditions for the experiment. We recommend the following:

   (a) Use a $20\times$ objective for the SEZ cultures.

   (b) Phase contrast images captured every cycle (we set 2–5 min as the duration of one cycle) (Note 3).

   (c) Select the total number of cell cycles to define the duration of the experiment. Typically the experiment requires between 5 and 7 days to record the lineage progression of aNSCs.

   (d) Use JPG or TIFF as exporting format.

4. Select the different positions (saved as x- and y-coordinates) and the focus (z-coordinate). Regions of the plate with the highest phase contrast conditions should be set in order to optimize the quality of the images. Define the exposure time for each position.

5. Subsequently, save these settings and run the experiment. Typically due to the small variations in temperature conditions that can occur during the experiment, changes in the focus conditions can happen. It is especially important to check the experiment periodically and refocus whenever it is required (Note 4).

3.3 Post-imaging Immunocytochemistry

1. Once the experiment is terminated, pause the experiment and remove the plate to perform immunocytochemistry in order to identify the progeny of the cells recorded during the experiment.

2. Aspirate and discard the culture medium and wash the cells twice with PBS.

3. Fix the cells for 10–15 min at room temperature by adding 400 µl of 4 % (wt/vol) PFA in PBS.

4. Wash three times with PBS and add the blocking solution (400 µl of PBS containing 2 % (wt/vol) BSA and 0.2 % (vol/vol) Triton X-100). Incubate for 1 h at room temperature (RT).

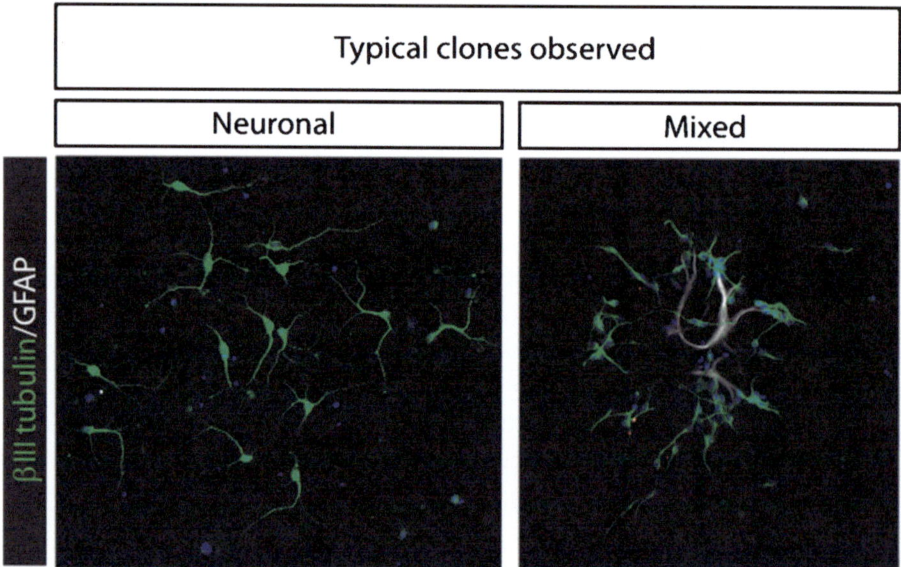

Fig. 3 Examples of clones derived from a single adult NSC. Cells in a culture of adult subependymal cells after 7 days in vitro, immunostained with antibodies against the neuronal marker βIII-tubulin (*green*) and the astrocyte marker GFAP (*white*). Observe the existence of clones containing only neurons (pure neuronal) and neurons alongside with astrocytes (mixed)

5. Remove the blocking solution and add the antibodies containing solution (antibodies diluted in 400 μl of PBS containing 2 % (wt/vol) BSA and 0.2 % (vol/vol) Triton X-100). Incubate then for 2 h at RT.

6. Wash three times with PBS (5–10 min each time) and add the secondary antibodies containing solution (antibodies diluted in 400 μl of PBS containing 2 % (wt/vol) BSA and 0.2 % (vol/vol) Triton X-100). Incubate then for 2 h at RT in the dark.

7. Wash three times with PBS. Cells should be kept in PBS for the subsequent steps.

8. Insert the plate back into the microscope device. Using the mark performed in Section 3.2, step 2, reset the *xyz* zero-positions in order to recover the precise coordinates of the different positions (Note 5).

9. Take pictures of each position using the proper fluorescence settings in order to reveal the cell-specific marker expression by immunofluorescence (Fig. 3).

10. Once the immunofluorescence pictures are taken, close the experiment and collect all the data produced during the experiment to proceed to the data analysis.

3.4 Data Processing and Analysis

1. Run the TTTlogfileconverter program and convert the images taken during the video time-lapse microscopy experiment into the format supported by the TTT software. All the instructions regarding hardware-operating system requirements, installation, and use of the TTT tracking software and TTTlogfileconverter software are available for download at http://www.helmholtz-muenchen.de/isf/haematopoese/software-download/index.html.

2. Once the software is properly installed run the TTT program, and select and load a position of the experiment to initiate the tracking. A window displaying every position used in the experiment associated to its number will be always available after running the TTT software.

3. The channels used during the video time-lapse microscopy experiment will be also displayed in the tracking window, 0 corresponding to phase contrast, 1, 2, etc. for the different fluorescence settings. Adjust the brightness and contrast for every channel.

4. Following the instructions described in the online handbook, track every cell selecting, when required, the corresponding option for each cell event (cell division, apoptosis, and lost cell). The lineage tree will be drawn automatically by the software in a different window as the tracking is progressing.

5. Once the lineage tree is tracked, match the last phase contrast live image with the fluorescence pictures acquired following post-imaging immunocytochemistry. This will allow for the identification of the progeny tracked during the experiment.

6. Finally, export the results (cell information, images, and lineage trees) into a new folder. The data can be used for statistic analysis and images can be assembled into movies by using adequate movie editors like ImageJ, which can be downloaded at (http://rsbweb.nih.gov/ij/).

4 Notes

The culture preparation and the following video time-lapse microscopy and data analysis comprise some critical steps that need special care to avoid negative results.

1. Isolating the tissue is a crucial phase in the protocol. Scratching the surface of the ventricle with the forceps or scissors can lead to damage or loss of the aNSC population located just beneath the surface (composed by the ependymal layer). Damage in the tissue could affect the amount of aNSCs isolated, recovering not enough cells and resulting in a suboptimal culture. Likewise, if the tissue layer containing the SEZ is too thick, it may contain

a high percentage of striatal tissue, increasing the quantity of dead cells and the content of myelin and thus leading to an excess of debris in culture. Excess of debris can compromise the survival of aNSCs and interfere with imaging.

2. This culture model is characterized by the absence of mitogenic factors; therefore the preparation should be performed carefully to avoid cell death. The main reasons that produce massive cell death after plating are listed below:

   (a) Excess of trituration during the dissociating step. As we previously mentioned in the Section 3.1, step 4, complete dissociation of the tissue is not required and the trituration should be interrupted as soon as the dissociating medium becomes turbid.

   (b) The overtime in the enzymatic digestion with trypsin can damage the cells leading to apoptosis. It is especially important that this step never exceeds the 30 min indicated in the protocol.

   (c) The culture medium should be fresh. Nonoptimal conditions in the culture medium could also result in cell death. If this happens we recommend preparing new medium or changing the batch of the B27 supplement.

   (d) This culture model is particularly sensitive to the time employed for the preparation. The performance of the culture should never exceed a total time of 3 h in order to avoid cell damage.

   (e) Cell death could also be a consequence of a poor cell adherence. An optimal coating is required for the adequate attachment of the aNSCs. If this problem is detected, fresh coating solution should be prepared.

   (f) If cell death is detected after placing the plate in the microscope device for time-lapse, this could indicate that the atmosphere conditions are not stable or are not set properly. Users should ensure that the temperature is 37 °C and CO_2 is at 8 %. Likewise the chamber surrounding the microscope stage must be properly closed to avoid variations that could compromise the culture viability.

3. Settings employed for image acquisition are also important for cell tracking and must be carefully selected. If the phase contrast images are acquired with low frequency the reconstruction of the lineage trees can be complicated due to an insufficient temporal resolution. On the other hand, a high frequency in phase contrast or fluorescent cycles can lead to a light-induced toxicity. In summary, cycles of acquisition should be balanced and shutters for both epifluorescence and transmitted light must be adjusted to the closed position between cycles.

4. Image quality is crucial for the subsequent cell tracking and preparation of the movies. Small variations in ambient conditions could affect the focus reducing the resolution of images. To avoid this problem we recommend checking and adjusting periodically the focus conditions as well as maintaining the microscope device and the experimental room under stable conditions. Ideally dedicate a room for the setup and avoid frequent entry.

5. Finally a mismatch between the last phase contrast live picture and the post-imaging immunofluorescence pictures can prevent the correct identification of the nature of the progeny. This problem is frequently produced by the incorrect resetting of the *xyz* zero-position. Reset the *xyz*-coordinates using the original mark at the bottom of the plate with the highest accuracy possible.

References

1. Zhao C, Deng W, Gage FH (2008) Mechanisms and functional implications of adult neurogenesis. Cell 132:645–660

2. Alonso M, Ortega-Perez I, Grubb MS et al (2008) Turning astrocytes from the rostral migratory stream into neurons: a role for the olfactory sensory organ. J Neurosci 28:11089–11102

3. Merkle FT, Mirzadeh Z, Alvarez-Buylla A (2007) Mosaic organization of neural stem cells in the adult brain. Science 317:381–384

4. Brill MS, Snapyan M, Wohlfrom H et al (2008) A dlx2- and pax6-dependent transcriptional code for periglomerular neuron specification in the adult olfactory bulb. J Neurosci 28:6439–6452

5. Brill MS, Ninkovic J, Winpenny E et al (2009) Adult generation of glutamatergic olfactory bulb interneurons. Nat Neurosci 12:1524–1533

6. Carleton A, Petreanu LT, Lansford R et al (2003) Becoming a new neuron in the adult olfactory bulb. Nat Neurosci 6:507–518

7. Hack MA, Saghatelyan A, de Chevigny A et al (2005) Neuronal fate determinants of adult olfactory bulb neurogenesis. Nat Neurosci 8:865–872

8. Lledo PM, Merkle FT, Alvarez-Buylla A (2008) Origin and function of olfactory bulb interneuron diversity. Trends Neurosci 31:392–400

9. Rietze RL, Reynolds BA (2006) Neural stem cell isolation and characterization. Methods Enzymol 419:3–23

10. Chojnacki A, Weiss S (2008) Production of neurons, astrocytes and oligodendrocytes from mammalian CNS stem cells. Nat Protoc 3:935–940

11. Reynolds BA, Weiss S (1992) Generation of neurons and astrocytes from isolated cells of the adult mammalian central nervous system. Science 255:1707–1710

12. Reynolds BA, Rietze RL (2005) Neural stem cells and neurospheres—re-evaluating the relationship. Nat Methods 2:333–336

13. Doetsch F, Petreanu L, Caille I et al (2002) EGF converts transit-amplifying neurogenic precursors in the adult brain into multipotent stem cells. Neuron 36:1021–1034

14. Golmohammadi MG, Blackmore DG, Large B et al (2008) Comparative analysis of the frequency and distribution of stem and progenitor cells in the adult mouse brain. Stem Cells 26:979–987

15. Luskin MB (1993) Restricted proliferation and migration of postnatally generated neurons derived from the forebrain subventricular zone. Neuron 11:173–189

16. Kuhn HG, Winkler J, Kempermann G et al (1997) Epidermal growth factor and fibroblast growth factor-2 have different effects on neural progenitors in the adult rat brain. J Neurosci 17:5820–5829

17. Lim DA, Alvarez-Buylla A (1999) Interaction between astrocytes and adult subventricular zone precursors stimulates neurogenesis. Proc Natl Acad Sci U S A 96:7526–7531

18. Scheffler B, Walton NM, Lin DD et al (2005) Phenotypic and functional characterization of adult brain neuropoiesis. Proc Natl Acad Sci U S A 102:9353–9358

19. Costa MR, Ortega F, Brill MS et al (2011) Continuous live imaging of adult neural stem cell division and lineage progression in vitro. Development 138:1057–1068

Methods Molecular Biology (2013) 1052: 13–28
DOI 10.1007/7651_2013_21
© Springer Science+Business Media New York 2013
Published online: 3 May 2013

Labeling and Tracking of Human Mesenchymal Stem Cells Using Near-Infrared Technology

Marie-Therese Armentero, Patrizia Bossolasco, and Lidia Cova

Abstract

The recently developed near-infrared (NIR) light imaging technology combines low background noise with deep tissue penetration and readily allows imaging and tracking of NIR-labeled cells, following transplantation in small animal model of diseases. The real-time and longitudinal detection of grafted cells in vivo, as well as their rapid ex vivo localization, may further clarify graft interactions with the surrounding, in target and nontarget organs throughout the body, over time. The present chapter describes a protocol for (1) the efficient labeling of human mesenchymal stem cells (hMSCs) using a membrane intercalating dye, emitting in the NIR 815 nm spectrum; (2) the stereotaxic transplantation of NIR815-hMSCs in rodent model of Parkinson's disease; and (3) the longitudinal in vivo detection of the grafted cells and the subsequent ex vivo imaging in selected tissues.

Keywords: Cell imaging, 6-Hydroxydopamine, Intra-striatal injection, Neurodegeneration, Stem cell transplantation, Cell therapy, Parkinson's disease

1 Introduction

Endogenous regeneration in the central nervous system (CNS) is very limited and allogeneic stem cells (SCs) have been increasingly recognized as a potential tool to replace or support lost cells injured by neurodegenerative processes. Several clinical studies, despite comparable setups, have reported contradictory results (1). Efficient SC therapy requires a still missing extensive comprehension of the progressive environmental cues and reciprocal interactions between the host and the donor cells. Facilitating and improving the number of cells engrafting in affected brain area are very important to achieve the best functional efficacy. This may be addressed by experimental work seeking to optimize cell delivery, improve characteristics of grafted cell, or modulate the CNS surroundings in the constant effort to better monitor cells and translate potential neurorescue mechanisms into novel targets for therapy (2).

The availability of experimental animal models to assess new therapeutic strategies in neurodegenerative disorders, including Parkinson's disease (PD), has been a fundamental breakthrough in the field of neuroscience research. Despite the increasing availability of transgenic and innovative models, the neurotoxin 6-hydroxydopamine (6-OHDA) remains one of the most widely used tools to induce a lesion in the rat (3). In particular, the intrastriatal (IS) injection of 6-OHDA in rats induces slow and retrograde degeneration of the nigro-striatal pathway that mimics, at least in part, the disease progression in humans. The characteristic progressive neuronal loss induced by this toxin in animals creates a therapeutic time-window wherein the neuroprotective potential of numerous pharmacological and non-pharmacological therapies has been successfully evaluated (3).

Near-infrared (NIR) light imaging offers new opportunities as a sensitive and noninvasive detection technique for diagnostic purposes. It represents an important advantage compared to other procedures, and avoids the use of retro- or lentiviruses, not transferable to clinical use. Particularly, it offers promising opportunities to develop noninvasive imaging protocols readily applicable to patients. This technology holds enormous potential for a wide variety of in vivo applications and is being increasingly used in small animal research. The use of NIR wavelengths for imaging allows deep penetration into tissues with minimal background and high optical contrast (4, 5). This simple, noninvasive technology consents live and real-time determination and imaging of biological targets without the need of exhaustive tissue sampling (6). In humans, the use of noninvasive NIR imaging has already been proposed as a routine diagnosis tool in stroke (7) and it is currently employed as a valuable bedside device for in vivo targeting of cancer and other tissue abnormalities (7, 8). Moreover, we have recently demonstrated that NIR technology allows longitudinal detection of fluorescent-tagged cells in living animals, giving immediate information on how different delivery routes affect cell permanence/persistence (9). NIR imaging allows investigations of transplanted cells from whole animals to the single-cell level over time, thus allowing cell tracking and assessment of integration/localization/migration in host surroundings and considerably reducing the number of animal experiments needed, as well as interindividual variability.

NIR technology can be readily accessible in any laboratory without the requirement for expensive clinical diagnostic equipment and specialized technical abilities, compared to those required for other in vivo imaging procedures (i.e., magnetic resonance). In this protocol we describe how human mesenchymal stem cells (hMSCs) can be rapidly and simply labeled with an NIR membrane intercalating dye, easily visualized in live animals after grafting in an animal model of PD and subsequently analyzed ex vivo using a suitable imaging platform.

2 Materials

2.1 Cell Labeling

1. Fully equipped sterile hood.
2. Water bath reaching 37 °C temperature.
3. Disposable sterile pipettes, flasks, polypropylene test tubes, and tips.
4. Centrifuge.
5. Human commercial mesenchymal stem cells (Cambrex, Walkersville, MD, USA).
6. Mesenchymal stem cell growth medium (MSCGM) (Cambrex).
7. Trypan blue (Invitrogen, Carlsbad, CA, USA).
8. Phosphate buffered saline (PBS) (without Ca^{2+} and Mg^{2+}) pH 7.2–7.4 (Invitrogen).
9. Trypsin 0.05 %–EDTA 0.02 % (Sigma-Aldrich, Saint Louis, MO, USA).
10. Fetal bovine serum (FBS) (Sigma-Aldrich).
11. CellVue® NIR815 Midi Kit for membrane labeling containing diluent C and dye (Polyscience, Warrington, PA, USA) (see Note 1).
12. Cytocentrifuge equipped with appropriate cuvettes and filters (Shandon, Pittsburgh, PA, USA).
13. Odyssey® Infrared Imaging System (LI-COR Biosciences, Lincoln, NE, USA).

2.2 Transplantation of Labeled hMSCs

1. Sprague–Dawley male rats (200 g at the time of surgery; Charles River, Calco, Italy).
2. Thiopental (50 mg/kg) (Rotexmedica GmbH, Trittau, Germany).
3. Sterile saline solution.
4. Gel foam (Harvard Apparatus Inc., Holliston, MA, USA).
5. Stereotaxic instrument for rat complete with ear bars (Stoelting, Kiel, WI, USA).
6. Optical fibers.
7. Disinfectant (any type is suitable).
8. Hamilton syringe 10 μl attached with a 26 gauge needle (7000 Series, Sigma-Aldrich).
9. Surgical tools: Scissors, sharp forceps, scalpels, sterile cotton pads, syringe and needles, metal clips, applying forceps.
10. Electric shaver.
11. Heating pad.
12. Odyssey® Infrared Imaging System (LI-COR).
13. MousePOD® in vivo Imaging Accessory (LI-COR).

2.3 Ex Vivo Analysis of Transplanted Labeled hMSCs

1. Cryostat.
2. Polylysine slides (Thermoscientific, Braunschweig, Germany).
3. Coverslides.
4. Pap pen (Sigma-Aldrich).
5. Neutral buffered formalin (NBF; Carlo Erba, Italy).
6. Primary antibody: Mouse anti-tyrosine hydroxylase, TH, (MAB 318, Millipore, Billerica, MA, USA).
7. Secondary antibody: IRDye® 700 goat anti-mouse antibody (LI-COR).
8. PBS (Invitrogen).
9. Triton X-100 (Sigma-Aldrich).
10. Tween 20 (Sigma-Aldrich).
11. Normal goat serum (NGS) (Sigma-Aldrich).
12. Blocking solution: PBS containing 10 % NGS and 0.3 % Triton X-100 (store fresh solution at 4 °C for up to 1 week).
13. Antibody solution: PBS containing 1 % NGS and 0.3 % Triton X-100 (store fresh solution at 4 °C for up to 1 week).
14. Odyssey® Infrared Imaging System (LI-COR).

3 Methods

3.1 Thawing and Culture of hMSCs

Commercial mesenchymal stem cells (Cambrex) are seeded following the manufacturer's instructions (see Note 2):

1. Quickly thaw cell-containing cryovials in a 37 °C water bath until the last sliver of ice melts.
2. Gently add thawed cells to a tube containing 5 ml of temperature-equilibrated MSCGM using a micropipette.
3. Centrifuge cell suspension at $200 \times g$ for 10 min at room temperature.
4. Count the total number of viable cells with the trypan blue exclusion method and seed them at a density of 5,300 cells per cm^2 (around 400,000 cells in a T75 flask).
5. Grow cells in an incubator at 37 °C, 5 % CO_2 for at least two passages (see Note 3).

3.2 NIR Labeling of hMSCs

Use cells at low passages (up to the fourth passage) (see Note 4). All the manipulations should be done at room temperature and preferably avoiding direct light exposure.

1. Discard supernatant from cell cultures and wash once with PBS.

2. Aspirate PBS, add 2 ml of trypsin–EDTA, and incubate at 37 °C, 5 % CO_2 for 5–10 min.

3. Gently shake the flask to resuspend loosely attached cells.

4. Add 10 ml of MSCGM to inactivate the trypsin–EDTA.

5. Collect cells in a 15 ml conical polypropylene tube and centrifuge at $500 \times g$ for 10 min at room temperature.

6. Wash cells with PBS: Resuspend cells in 10 ml PBS and centrifuge as above (step 5).

7. Resuspend cell pellet in a small volume of PBS (around 100 μl) and transfer to a small conical bottom 1.5 ml polypropylene tube.

8. Centrifuge as above (step 5) and wash once again with PBS (volume around 100 μl).

9. After centrifugation, carefully aspirate the supernatant with a micropipette or an insulin syringe, being careful not to remove any cells.

10. Resuspend cells in 50 μl of diluent C, contained in the CellVue 97® NIR815 Midi Kit (for 1×10^6 cells) with gentle mixing (see Note 5).

11. Immediately prior to staining prepare the staining solution by mixing 50 μl of diluent C and 1 μl of NIR815 dye in a sterile 200 μl polypropylene tube (see Note 6).

12. Add the staining solution to the cell suspension, immediately mix the sample by pipetting, and incubate for 5 min with periodic mixing (see Note 7).

13. Block staining by adding 200 μl of FBS, transfer cells in 15 ml conical bottom polypropylene tube, and wash cells twice with 10 ml of MSCGM to eliminate unbound dye.

14. Resuspend cells in MSCGM (for 1×10^6 cells use 500 μl) in a sterile 500 μl polypropylene tube. Keep a small aliquot (5 μl) to assess adequate labeling of cells (see Section 3.3).

15. Use 10 μl of the cell suspension to check viability of the cells by trypan blue exclusion test.

16. Plate cells in a T75 cm^2 flask in MSCGM overnight, if needed (see Note 8).

3.3 Assessment/ Control of Adequate NIR815 Cell Labeling

To check the adequate labeling of cells, a cytocentrifugation onto glass slides, immediately after staining procedures, may be performed.

1. Place slides and filters into appropriate slots in the cytospin and load not fewer than 100 μl of cell suspension (containing around 20,000–50,000 hMSCs) in each cuvette (see Note 9).

2. Spin the sample for 7 min at 400 rpm (see Note 10).

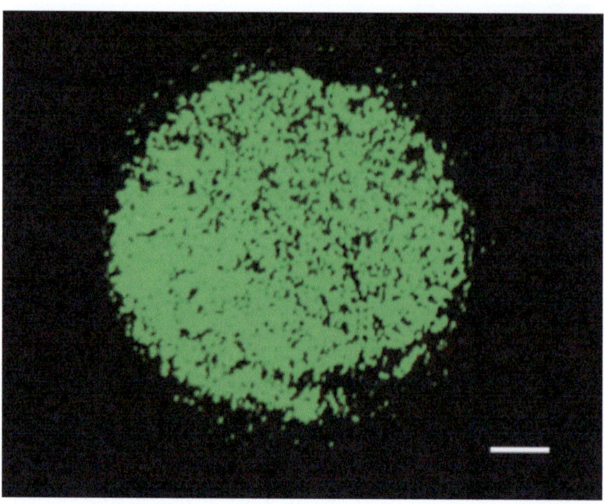

Fig. 1 Cytospin. Representative image showing a cytospin of NIR815-labeled hMSCs. Scale bar: 1 mm. (Modifed from Bossolasco et al., Int J Nanomed 2012)

3. Remove the filters from their slides and proceed to scan using the Odyssey® Infrared Imaging System.

4. Place the slide on the surface of the imager cells facing upwards and determine the scan area.

5. Using the software dedicated to the imager insert the coordinates of the scan area and set the scan parameters. Set the "offset focus" to 1 mm (see Note 11). At this point a rapid scan with low resolution (169 μm) and quality (lowest) is enough to assert adequate labeling of cells (Fig. 1). Scan intensity is usually set to 3 but needs to be adapted to the amount of cells (see Note 12).

3.4 Intra-cerebral Transplantation of NIR815-Labeled hMSCs

All animal procedures are carried out in accordance with the European Communities Council Directive of November 24th, 1986 (86/609/EEC), and are approved by the local Animal Care Committee.

The original transplantation protocol has been applied to PD-like rats bearing a unilateral 6-OHDA-induced lesion of the nigro-striatal pathway already described elsewhere (9). The transplantation procedure described below can however be applied to any animal model of neurodegeneration or brain injury (see Note 13).

1. Prepare the stereotaxic frame mounted with the Hamilton syringe.

2. Weigh and anesthetize male Sprague–Dawley rats using 50 mg/kg thiopental (intraperitoneal injection). Leave the animal in a cage with bedding until unconscious (see Note 14).

3. Accurately shave the animal's head with an electric shaver to uncover skin and wipe the area with a sterile cotton pad imbedded with disinfectant solution.

4. Place the anesthetized animal in the stereotaxic frame with the tooth bar set at -3.3 mm (see Note 15).

5. Perform a net 2 cm midline incision on the skin with a sharp scalpel starting between the eyes and fix the skin with two metal clamps to keep the skull uncovered.

6. Remove the connective tissue above the bone by scraping with the scalpel to allow visualization of bregma.

7. Find the bregma and point the needle exactly on it. Read the precise anterio-posterior (AP) and lateral (L) coordinates on the manipulator (x- and y-axis) of stereotaxic frame and retract the needle. Calculate the location for the injection site and then move and lower the needle to the specific coordinates (see Note 16).

8. Evidence the injection position on the skull with a fine marker and retract the needle.

9. Drill a small hole through the skull bone above the injection coordinate until you reach the dura mater (see Note 17).

10. Fill the syringe with the previously optimized cell dose (see Note 18).

11. Lower the needle until it reaches the dura mater without piercing it (see Note 19).

12. Determine the DV position on the stereotaxic frame. Calculate the injection coordinate and slowly lower the needle to the desired depth (see Note 20).

13. Slowly infuse the cells (see Note 21). Cells are injected at 0.5 μl/min and needle is left in place to allow diffusion of the injected volume.

14. After at least 5 min retract the needle VERY SLOWLY to avoid suction of the cells along the needle track due to pressure.

15. Fill burr hole with foam and clip wound.

16. Place the animal in a clean cage with heating pad until it regains consciousness.

3.5 In Vivo Tracking of NIR-Labeled hMSCs

1. In vivo tracking of NIR815-labeled cells can be performed at any time point following intracerebral transplantation. Before starting with the animal procedure turn on the Odyssey® Imager (LI-COR) and MousePOD® in vivo Imaging Accessory (Fig. 2a) so that the chamber can reach the desired temperature (37 °C). Constant temperature is maintained in the chamber throughout the in vivo imaging procedure.

2. Weigh and anesthetize transplanted rats using thiopental (50 mg/kg, intraperitoneal injection). Animals need to be

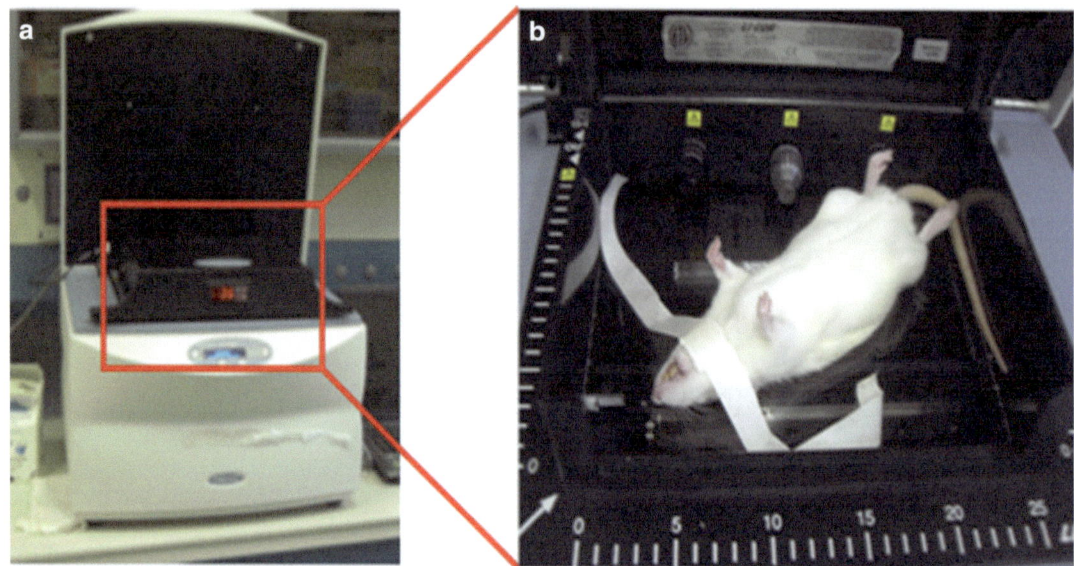

Fig. 2 NIR imaging system. (**a**) Photograph of the Odyssey® Infrared Imaging System mounted with MousePOD® in vivo Imaging Accessory. (**b**) Photograph of anesthetized animal adequately positioned on the glass surface of the imager to allow correct imaging procedure

maintained under sedation throughout the entire in vivo imaging procedure (see Note 22).

3. Accurately shave the fur of the animal in the area that will be exposed to the imager (see Note 23).

4. Place the animal on the instrument in a supine position so that the area to be scanned is in direct contact and lies flat with the imager surface, determine the scan area, and close the chamber (Fig. 2b) (see Note 24).

5. Insert the coordinates of the scan area, set the scan parameters in the dedicated software, and initially perform a rapid scan at low resolution (169 μm) and quality (lowest) to assert the precise localization of the transplanted cells and determine the best "focus offset." Scan intensity of the 800 nm laser is typically set between 7 and 10 (see Note 25).

6. Perform a second scan to obtain a high-resolution image (Fig. 3a) (see Note 26).

7. Once the scan is completed a semiquantitative evaluation of the NIR815 can be performed (Fig. 3b) (see Note 27).

8. Images can be transformed in pseudo-colors to obtain a visual representation of NIR815 intensity levels (Fig. 3c).

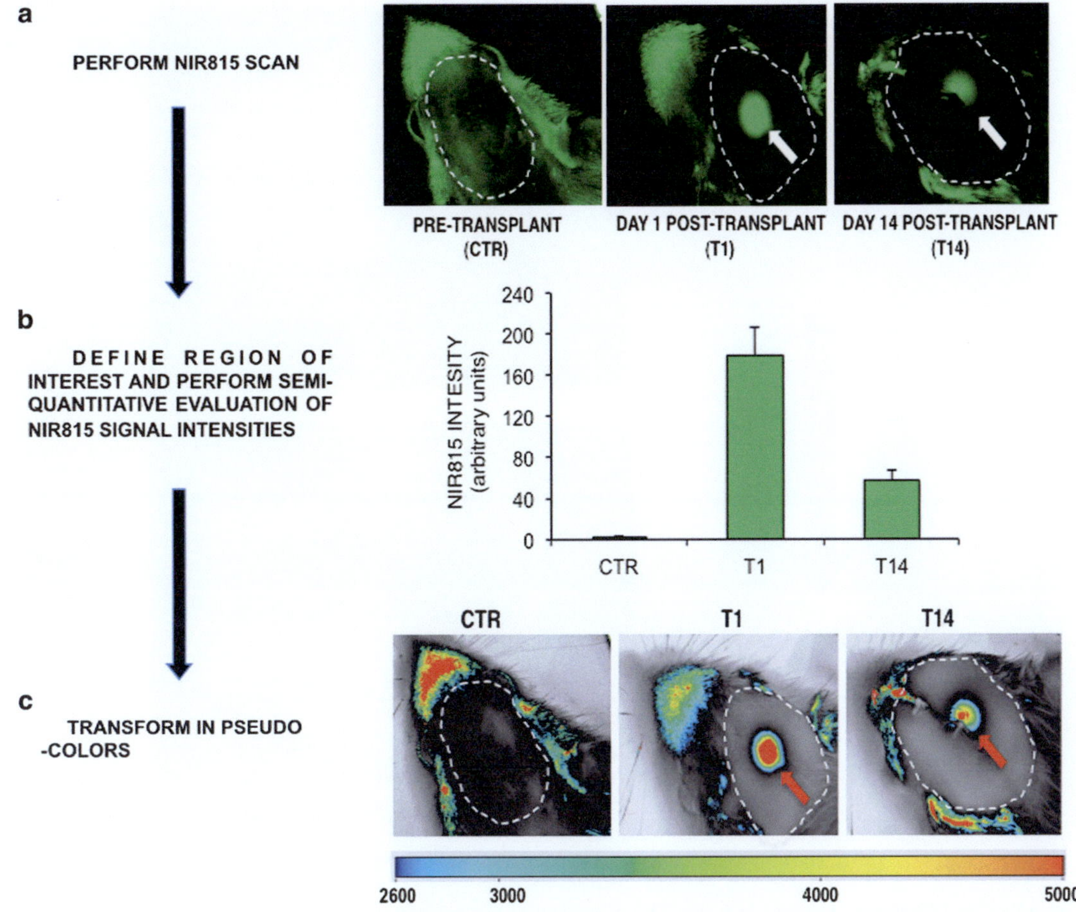

Fig. 3 In vivo and ex vivo NIR imaging of grafted NIR815 hMSCs. (a) Representative photograph of longitudinal NIR imaging in a live animal. *Dotted circles* represent the area of interest selected for the analysis. (b) Semiquantitative analysis of the scan. (c) Transformation of NIR815 intensities in pseudo-colors. (Modified from Bossolasco et al., Int J Nanomed 2012)

3.6 Ex Vivo Imaging of NIR-Labeled hMSCs

3.6.1 Whole Organ

1. At the end of the experimental paradigm anesthetize the animal (100 mg/kg thiopental) and sacrifice it by decapitation (see Note 28).

2. Immediately remove brain, place it on a glass slide, and scan rapidly. (Note that to avoid degradation of the tissue only a low-resolution (169 µm) and low-quality scan (lowest) is performed. The green signal indicates the presence of the transplanted cells (Fig. 4).) Moreover, at the time of sacrifice any organ may be removed and scanned to evaluate potential migration of cells to peripheral tissues/organs (see Note 29).

3. Freeze organs on dry ice immediately after scanning and keep at −80 °C for future evaluation.

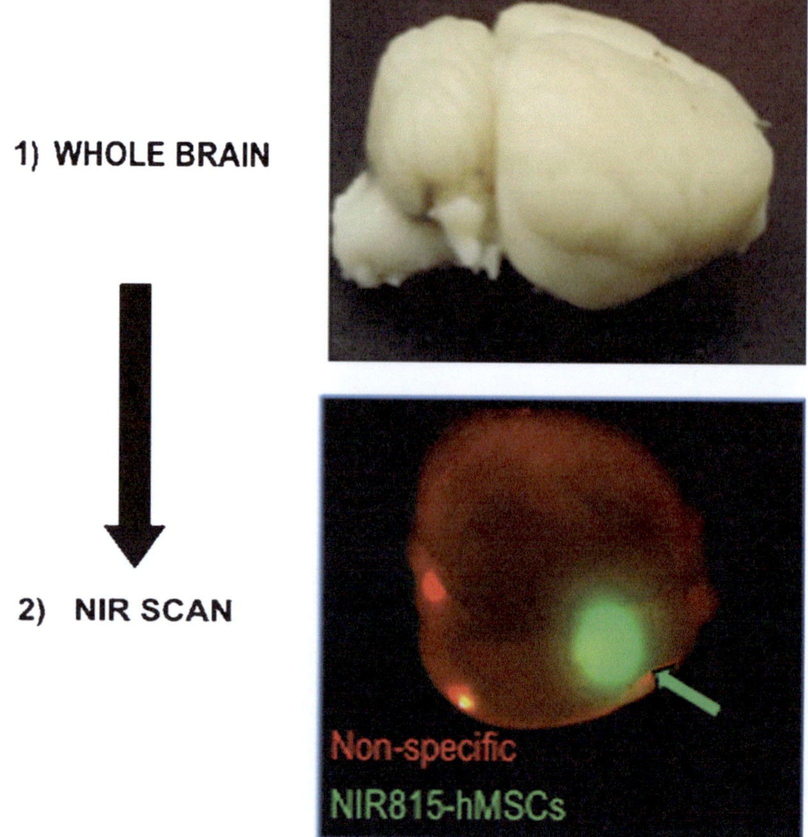

Fig. 4 Ex vivo NIR imaging of whole brain. The presence of the grafted cells is clearly indicated by the intense NIR815 signal (*green arrow*) visible at the site of transplantation. The *red* signal corresponds to a nonspecific background of the tissue in the *red* spectrum. (Modified from Bossolasco et al., Int J Nanomed 2012)

3.6.2 Tissue Sections

1. Cut brain coronal sections (25 μm) using a cryostat and mount them on polylysine slides. Up to four sections containing the striatum can be mounted on each slide.

2. Let the sections dry, protected from light, for at least 30 min.

3. Place all slides side by side on imager surface, sections facing upwards.

4. Determine and insert the coordinates of the selected area and set the scan parameters to low resolution (169 μm) and low quality (lowest) to perform a high-throughput evaluation of all the brain sections (Fig. 5) (see Note 30).

5. Select slides and perform high-resolution scan, if necessary.

6. Selected slides can be processed for immunohistochemistry (Section 3.4, step 3).

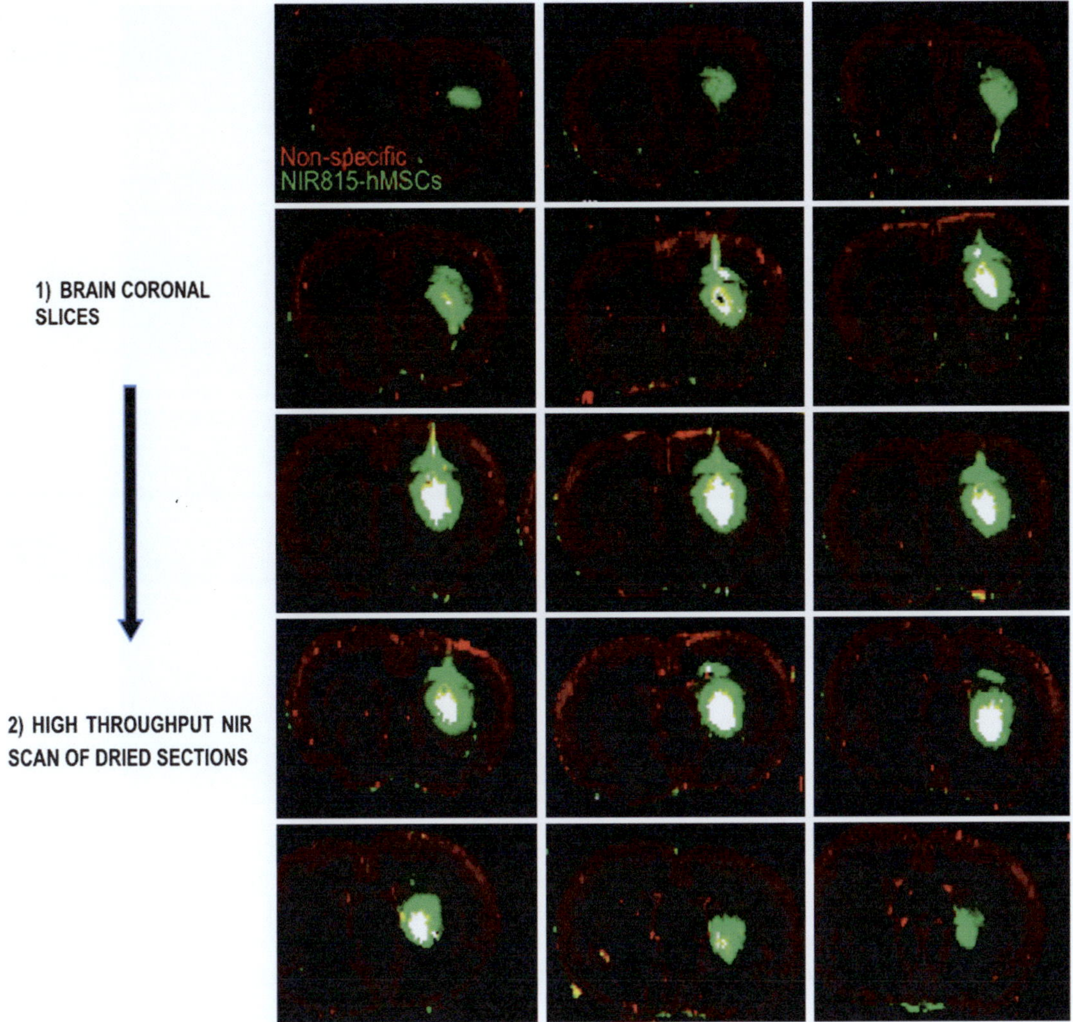

Fig. 5 High-throughput NIR imaging. Coronal brain sections containing the entire striatum are mounted on glass-slides and a high-throughput analysis can be performed at low resolution. Presence of the grafted hMSCs is clearly indicated by the intense *green* signal. The *red* signal corresponds to a nonspecific background of the tissue in the *red* spectrum

3.6.3 NIR Immunohistochemistry for Tyrosine Hydroxylase (TH)

Localization of transplanted NIR815-labeled cells within the 6-OHDA-lesioned striatum can be performed by conventional immunohistochemical procedure using secondary antibodies linked to an NIR fluorochrome (IRDye®).

1. Circle the sections with a pap pen.

2. Fix section for 15 min with NBF.

3. Wash with PBS for 10 min (see Note 31); repeat wash four times.

4. Block sections with the blocking solution for 1 h at room temperature.

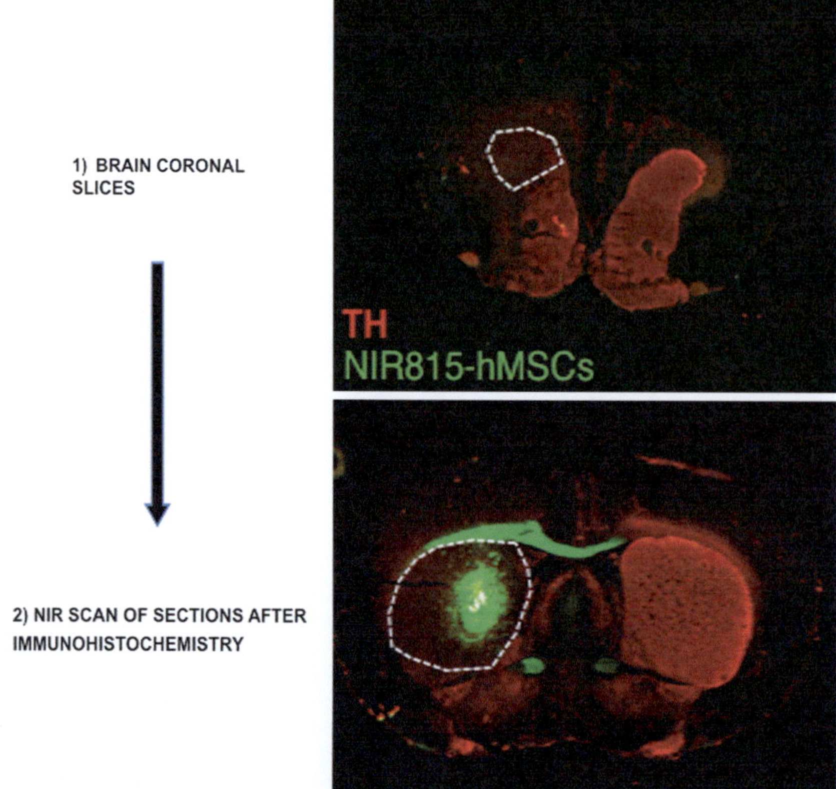

Fig. 6 NIR immunohistochemistry. A classical immunohistochemistry procedure is performed using a secondary antibody emitting in the *red* NIR spectrum (700 nm). At the end of the procedure sections are dried and can be scanned immediately. The *red* signal indicates the presence of tyrosine hydroxylase (TH)-positive dopaminergic terminals in the intact striatum. Localization of the grafted cells in the lesioned striatum (absence of *red* signal) is indicated by the intense *green* signal (800 nm). (Modified from Bossolasco et al., Int J Nanomed 2012)

5. Incubate with the primary anti-TH antibody in the antibody solution overnight at 4 °C (*see* **Note 32**).

6. Wash sections with PBS containing 0.1 % Tween 20 for 5 min; repeat wash four times.

7. Dilute secondary antibody (1:5,000) in the antibody solution and incubate for 1 h at room temperature.

8. Wash with PBS–0.1 % Tween 20 for 10 min; repeat wash six times.

9. Wash twice with PBS.

10. Wash once with distilled H_2O.

11. Let slides dry overnight protected from light (*see* **Note 33**). Perform high-resolution scan (*see* **Note 28**: sections do not need to be mounted to obtain high-quality NIR images) (Fig. 6).

4 Notes

1. The NIR815 dye emits in the red spectrum and results in higher signal specificity and reduced background noise compared to NIR700 dyes.

2. Any other adherent or non-adherent cell types may be used, but will require previous optimization of the protocol.

3. Divide cells when they reach confluence; this will depend on their specific growth rate.

4. A karyotyping of cells may be performed to assess the absence of cell culture-induced aberrations.

5. A minimum of 1×10^6 cells can be labeled with this procedure. The volume indicated is optimized for 1×10^6 cells. If the number of cells to be labeled is larger, the volume needs to be proportionally adapted (i.e., 250 µl for 5×10^6 cells, 500 µl for 1×10^7 cells, etc.).

6. The volume of staining solution needs to be equal to the volume of the cell suspension. Note that concentration of the NIR815 dye should not exceed 1 % of the final volume. Higher percentage may affect cell viability. Dye concentration should be optimized for each cell type.

7. Labeling times exceeding 5 min may affect cell viability. Incubation time should be optimized for each cell type.

8. Cells may be transplanted immediately after staining or, if needed, may be grown in culture overnight or more (be careful that, although cell labeling is stable over time, the dye will be diluted by cell division over passaging).

9. Dilute the aliquot with PBS to obtain a final volume of at least 100 µl.

10. Alternatively, a drop of cell suspension may be placed onto a glass slide, sealed with a coverslip, and directly observed on the Odyssey® Infrared Imaging System.

11. Focus will depend on the thickness of the slide. A rapid scan can be previously performed to determine the optimal offset focus.

12. If necessary high resolution (21 µm) and quality (highest) can be performed. Highest resolution and quality of the image will take longer time to complete the scanning procedure.

13. Protocol for 6-OHDA injection has been described in details elsewhere (10).

14. Any other anesthetic may be used.

15. Before fixing the teeth of the animal assure that the animal is correctly placed; no left–right movement of the head should be possible. Correcting placing of the animal is fundamental

to properly calculate the coordinates and find the correct injection site.

16. The site of injection is determined using a brain atlas specific for the species of interest. For rats we use the Paxinos and Watson brain atlas (11). Find the brain area of interest in the atlas and determine the coordinates. Injection coordinates are expressed in millimeters relative to bregma. For example for unilateral injection of hMSCs in the right striatum the needle has to be moved to the following position: $AP = +1$ mm and $L = -3$ mm from bregma.

17. Be very careful not to damage the membrane; it will be used to calculate the dorsoventral (DV) coordinates.

18. Cells need to be resuspended in physiological solution or PBS; maximal volume that may be injected without damaging neuronal structure has to be determined a priori and will depend on the injection site. In addition optimum cell concentration will vary with each cell type; perform preliminary trials to determine the maximum cell concentration that can be used and that will not cause tapping of the needle.

19. To precisely assess the point at which the needle touches the dura, place yourself at an angle with respect to the surface of the membrane so that you may observe the reflection of a light beam. Carefully and slowly lower the needle; as the needle touches the membrane, the latter will be displaced, thereby modifying reflection of the light beam that will no longer be visible from your viewpoint.

20. For a single IS injection $DV = -5$ mm with respect to dura: Global coordinates with respect to bregma and dura for single unilateral injection in the right striatum are $AP = +1$ mm, $L = -3$ mm, and $DV = -5$ mm.

21. Manually and slowly rotate the tip of the Hamilton plunger with a light downward pressure until the desired volume has been injected.

22. The MousePOD® in vivo Imaging Accessory can be connected to an anesthetic system. Refer to the Li-Cor Web site for major information (http://www.licor.com/bio/products/accessories/odyssey_accessories/mousepod/mousepod.jsp).

23. Complete removal of the fur is fundamental for NIR analysis; the animal's hair may act as an optic fiber, thus interfering with the imaging process.

24. To obtain a distinct/clear signal any possible movements of the animal's head should be avoided.

25. Best detection of cells transplanted in the striatum is usually obtained with focus set at 4 mm.

26. Resolution of 84 μm reduces time of scan and allows the acquisition of high-quality images. Set scan intensity between 7 and 10 for the 800 nm laser (green); if necessary a high, 21 μm resolution can be performed but time to complete the scan will be considerably extended; the 700 nm channel (red) is not necessary for detection of NIR815-labeled cells and is usually excluded.

27. Although NIR values can be precisely determined, measurements should be considered semiquantitative because they may be slightly influenced by the position of the animal (head) on the surface of the imager. For repeated evaluations over time, users are advised to document the original position and scan area of the animal during the first imaging procedure and conform to them as closely as possible during successive measures.

28. Transcardiac perfusion of the animal is not necessary for ex vivo analyses. If necessary for other purposes, it can be performed and will not interfere with subsequent imaging procedures.

29. Alternatively, brain can be frozen on dry ice and scanned afterward.

30. Scan intensity will depend on the amount of NIR815-labeled cells present in the sections and is usually set between 3 and 5. Focus is normally set to 1 mm (thickness of the slide). Complete scanning of 30 sections typically takes 10 min. This first scan will determine the specific localization of the transplanted cells in sections and allow selection of more accurate areas for the subsequent analysis.

31. Add enough PBS so that sections are completely covered; 200–300 μl are typically required for four sections; this holds true for all subsequent washes and incubations with primary and secondary antibodies.

32. Best dilution of the primary antibody has to be previously determined for each lot; we found that a 1:1,000 dilution usually gives optimal results.

33. For long-term storage slides can be mounted with Fluorosave® and covered with coverslips. The mounting solution does not interfere with the imaging procedure.

References

1. Lunn JS et al (2011) Stem cell technology for neurodegenerative diseases. Ann Neurol 70:353–361

2. Bhakoo K (2011) In vivo stem cell tracking in neurodegenerative therapies. Expert Opin Biol Ther 11:911–920

3. Blandini F, Armentero MT (2012) Animal models of Parkinson's disease. FEBS J 279:1156–1166

4. Aswathy RG et al (2010) Near-infrared quantum dots for deep tissue imaging. Anal Bioanal Chem 397:1417–1435

5. Luker GD, Luker KE (2008) Optical imaging: current applications and future directions. J Nucl Med 49:1–4

6. Kosaka N et al (2009) Clinical implications of near-infrared fluorescence imaging in cancer. Future Oncol 5:1501–1511

7. Lareau E et al (2011) Multichannel wearable system dedicated for simultaneous electroencephalographynear-infrared spectroscopy real-time data acquisitions. J Biomed Opt 16:096014

8. He X et al (2010) Near-infrared fluorescent nanoprobes for cancer molecular imaging: status and challenges. Trends Mol Med 16:574–583

9. Bossolasco P et al (2012) Noninvasive near-infrared live imaging of human adult mesenchymal stem cells transplanted in a rodent model of Parkinson's disease. Int J Nanomedicine 7:435–447

10. Mercanti G et al (2012) A 6-hydroxydopamine in vivo model of Parkinson's disease. In: Kurien BT, Scofield RH (eds) Methods in molecular biology. Humana, New York, pp 355–364

11. Paxinos G, Watson C (1998) The rat brain in stereotaxic coordinates. Academic, San Diego

Methods Molecular Biology (2013) 1052: 29–39
DOI 10.1007/7651_2013_25
© Springer Science+Business Media New York 2013
Published online: 3 May 2013

High-Content Imaging and Analysis of Pluripotent Stem Cell-Derived Cardiomyocytes

Gábor Földes and Maxime Mioulane

Abstract

Human pluripotent stem cells (hPSC) are investigated as a source of authentic human cardiac cells for drug discovery and toxicological tests. Cell-based assays using automated fluorescence imaging platform and high-content analysis characterize hypertrophic and toxicity profiles of compounds in hPSC-derived cardiomyocytes (hPSC-CM) at the cellular and subcellular levels. In purified population of hPSC-CM loaded with cell tracer probe and cell death markers, both hypertrophic and toxicity profiles can be assessed in live cardiomyocyte cultures. Alternatively, in non-purified cultures of hPSC-CM, hypertrophy, proliferation, and cell death assays can be performed specifically in the cardiomyocyte subpopulation using antibodies directed against cardiac proteins and a combination of cell death- and proliferation-specific fluorescent probes.

Keywords: Human pluripotent stem cell, Cardiomyocyte, Hypertrophy, Cell death, Proliferation, Automated high-content imaging

1 Introduction

Human pluripotent stem cell-derived cardiomyocytes (hPSC-CM) are being investigated as a new source of cardiac cells for drug safety assessment and disease modelling. Differentiation protocols for production of cardiomyocytes from hPSC, including embryonic and induced pluripotent stem cells, are now efficient, relying on sequential exposure to cardiogenic factors that mimic early cardiac development (1). Purified cardiomyocyte cultures can also be purchased from various commercial sources. Pharmaceutical companies are showing growing interest in using these cells for drug development and toxicology, hoping that these humanized platforms will both increase predictive capabilities and decrease drug development costs. There has also been uptake from academic scientists seeking an alternative to animal-derived cells (2). Here we

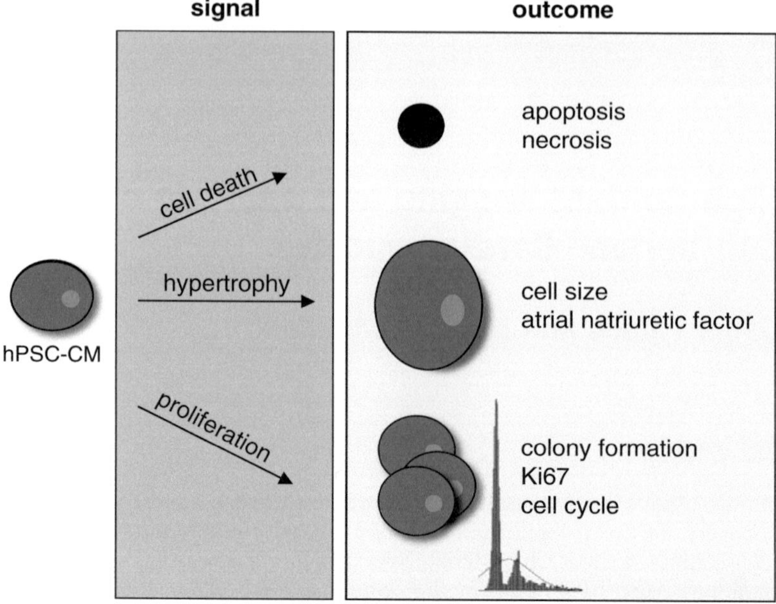

Fig. 1 Schematic drawing on the homeostasis of cultured human pluripotent stem cell-derived cardiomyocytes. Depending on the nature of the stimuli and their intensity, hPSC-CM can undergo hypertrophy, activate cell death pathways, or progress into cell division

describe a novel scalable high-content microscopy-based method for the detection of cell death, proliferation, and hypertrophy in hPSC-CM that can serve for predictive in vitro cardiomyocyte screens (Fig. 1).

2 Materials

2.1 Human Pluripotent Stem Cell-Derived Cardiomyocytes

1. Extracellular matrix Matrigel (BD Biosciences)-coated plates as substrates for feeder-independent human ESC or hiPSC cultures (see Note 1).

2. Predefined stem cell qualified medium such as Nutristem (Stemgent) for culturing undifferentiated stem cells.

3. Cytokines and medium for directed cardiac differentiation in monolayer: Human recombinant Activin A (final concentration: 100 ng/ml), and bone morphogenetic protein 4 (BMP4, 10 ng/ml). For stock, reconstitute Activin A at 50 μg/ml A in sterile PBS containing at least 0.1 % human or bovine serum albumin. Reconstitute BMP4 at 25 μg/ml in sterile 4 mM HCl containing at least 0.1 % human or bovine serum albumin. Culture medium for differentiation: RMPI medium with 2 % B27 (insulin-free) supplement (Invitrogen).

4. Serum-free basal medium for hypertrophy experiments: DMEM:M199 medium, in 3:1 ratio. Add 1 ml penicillin/streptomycin, 0.2 g bovine serum albumin (0.2 % wt/vol), 0.00176 g ascorbic acid, 0.066 g creatine, 0.0626 g taurine, 0.03224 g carnitine to 100 ml DMEM/M199 medium. Filter on 0.22 μm low protein binding acrodisk and keep at 4 °C. Dilute insulin (Sigma) 1:10 and then add 8.2 μl of this to 50 ml DMEM/M199 on the day of experiments.

5. Alternatively, hESC-CM and hiPSC-CM can be purchased from commercial suppliers (see Note 2).

2.2 Apoptotic and Cell Growth Agents

1. Use doxorubicin as a positive control apoptotic agent in cardiac cells. Prepare stocks at a concentration of 10 mM in DMSO and keep up aliquots up to 6 months at −20 °C. Make up diluted doxorubicin solutions in RPMI supplemented with B27 (containing insulin) (see Note 3).

2. Hypertrophic alpha-adrenoceptor agonist phenylephrine (10 μM). Use serum-free basal medium for dilution. Prepare fresh phenylephrine solution from powder before each experiment. Alternatively, treat cells with angiotensin II (100 nM) or endothelin-1 (1, 10, and 100 nM, all from Sigma) for 48 or 24 h, respectively.

2.3 Antibodies and Vital Dyes

1. Primary antibodies for immunocytochemistry: Anti-Ki67 (1:100), anti-atrial natriuretic factor (1:300, Abcam), anti-myosin heavy chain α/β (MHC α/β, clone 3-48, 1:200, Abcam), and anti-active caspase-3 (1:250, Abcam) primary antibodies.

2. Detection of primary antibodies: Alexa 488-, Alexa 647-, and Alexa 546-conjugated secondary antibodies (all 1:400, Invitrogen). For cell death assay using vital dyes, prepare various combinations of Hoechst 33342 (0.5 μg/ml), caspase-3&7 fluorescent substrate kit (Vybrant® FAM Caspase-3 and -7 Assay Kit, Invitrogen), TMRM (20 nM, Invitrogen) or Cell Tracker Red (5 mM, Invitrogen), and ToPRO-3 (1:1,000, Invitrogen) or BOBO-1 (1:1,000, Invitrogen). Dilute probes in RPMI medium supplemented with B27. See also Section 3.2.

2.4 Image Acquisition and Analysis

1. Inverted phase contrast microscope for testing quality and confluency of stem cell cultures (see Note 4).

2. Acquire images on automated microscopy and image analysis system with 10× objective suitable for filter sets.

3. Image analysis by using Cellomics Arrayscan HCS algorithms (Thermo Scientific).

3   Methods

3.1   Cardiac Differentiation

1. Culture undifferentiated human pluripotent stem cells on Matrigel-coated 6-well plates in 2 ml/well Nutristem medium (see Note 1). Change medium every day and subculture when the cells reach 80 % confluence. Check cultures under inverted phase contrast light microscope on a daily basis.

2. Perform cardiac differentiation using dense monolayers of hESC treated with 100 ng/ml Activin-A for 24 h and 10 ng/ml BMP4 for another 4 days in RPMI medium supplemented with insulin-free B27 (3). From day 5, feed cultures every other day with RPMI supplemented with B27 (containing insulin). First beating areas typically arise after 15 days following induction (see Note 5).

3. Isolate cardiomyocytes using trypsin–EDTA (Gibco) for 3–5 min and collagenase IV (200 U/ml) for another 10 min. Check cell culture under light microscope after 5 min. Triturate clusters enriched in cardiomyocytes with a micropipette, count cells using a *hemocytometer*, and plate 50,000 single cells/cm^2 onto 0.1 % gelatine-coated 96-well plates (see Note 6).

3.2   Live Cell Staining

1. To perform live staining, use pure cardiomyocyte populations (i.e., isolated rat neonatal ventricular myocytes and hiPSC-CM from Cellular Dynamics International).

2. Add drugs 3–5 days after cell plating or cell recovery from frozen stocks. Every well should contain the same amount of media (typically 100 μl).

3. After drug treatment, incubate cells with Cell Tracker Red for 45 min at 37 °C with 5 % CO_2. Remove the solution and replace with caspases-3 and -7 substrate solution for 60 min at 37 °C with 5 % CO_2. Rinse with Apoptosis Buffer (Invitrogen). Incubate cells with Hoechst 33342 and ToPRO-3 for another 10 min. Place cells in 100 μl fresh medium (RPMI with B27) in 96-well plate.

4. Alternatively, if hypertrophy readout is not required, cell death assay can be modified using TMRM instead of Cell Tracker Red. Add TMRM to the Hoechst/ToPRO-3 mixture and incubate cells for 10 min (see Note 7) (Fig. 2).

5. Place labelled cultures in the Cellomics chamber at 37 °C/5 % CO_2 and scan using automated microscope (see Section 3.4).

3.3   Immunocytochemistry

Use combinations of immunocytochemistry markers to characterize detailed hypertrophic and cell death properties of hPSC-CM cultures.

1. Fix the cells with 4 % paraformaldehyde for 10 min and wash three times with PBS.

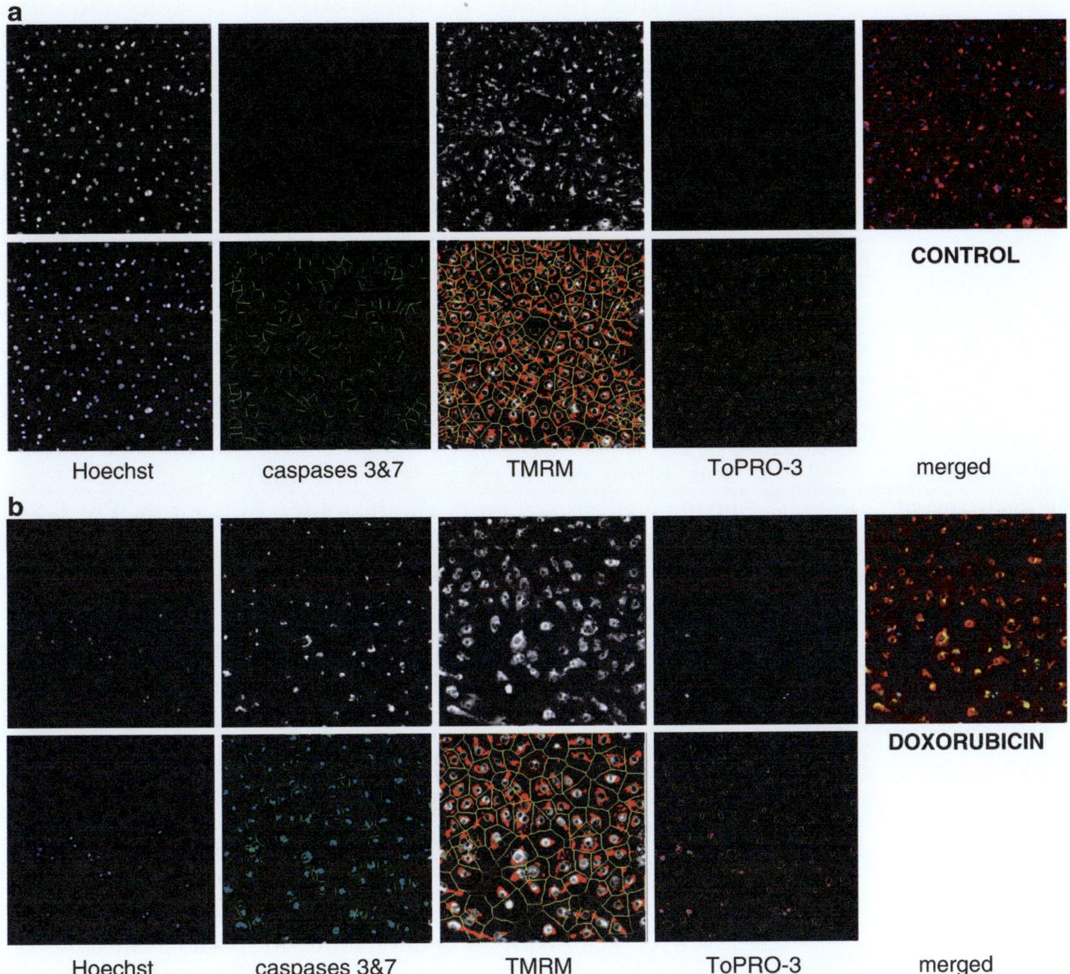

Fig. 2 Representative high-content images from doxorubicin-induced cell death profile in DMSO control (**a**) and doxorubicin-treated (**b**) hiPSC-CM cultures. Doxorubicin is a strong inducer of DNA degradation (Hoechst), and early apoptosis (caspase 3&7 substrate without cell membrane permeabilization) with intact mitochondrial membrane potential (TMRM). ToPRO-3 was used as necrosis marker

2. Permeabilize cells with 0.2 % Triton-X100 for 10 min, block with 4 % fetal bovine serum in PBS for 1 h, and incubate with primary and secondary antibodies (in 3 % bovine serum albumin in PBS as a carrier solution). After immunostaining, keep the cells in PBS and store plates at 4 °C.

3. For hypertrophy and proliferation assessments, rinse wells with PBS and fix plates. After permeabilization and blocking, label cells with anti-Ki67 (Abcam, 1:100), anti-atrial natriuretic factor (Santa Cruz, 1:300), and anti-MHC α/β (clone 3-48, Abcam, 1:200) primary antibodies and incubate cells for 1 h at room temperature. Wash plates three times with PBS

and detect primary antibodies with Alexa 488- and Alexa 568-conjugated secondary antibodies (all 1:400, 45 min). Add Hoechst for 10 min to visualize chromatin. Wash plates three times and add 100 μl PBS per well.

4. For cell death assay, incubate live cells with BOBO-1 and Hoechst 33342 for 10 min prior to fixation. Wash wells with PBS and fix plates. Permeabilization and blocking as described in Section 3.3, step 1. Incubate with rabbit polyclonal anti-caspase 3 (1:500), and mouse monoclonal anti-MHC α/β (1:200), for 1 h at room temperature. Wash plates three times with PBS and detect primary antibodies with Alexa 546- and Alexa 647-conjugated secondary antibodies (all 1:400, 45 min). Wash plates three times and add 100 μl PBS per well.

3.4 Algorithm for High-Content Imaging of Human Pluripotent Stem Cell-Derived Cardiomyocytes

1. Scan plates on ArrayScan™ VTi automated microscopy and image analysis system (Cellomics Inc., Pittsburgh, PA, USA). Adapt and use in-built Spot Detector, Target Activation, Cell Cycle, Morphology Explorer, and Compartmental Analysis BioApplication protocols.

2. Use the system of automated highly sensitive fluorescence microscope with $10\times$ objective and suitable filter sets, identify stained cells with Hoechst in fluorescence channel 1.

3. For hypertrophy assays detect MHC α/β-Alexa 488 in channel 2 and ANF- and Ki67-Alexa546 in channel 3, respectively.

4. For cell death/hypertrophy assays detect caspases 3&7 substrate (FITC) in channel 2, Cell Tracker or TMRM in channel 3 (TRITC), and ToPRO-3 in channel 4 (Cy5).

5. For fixed cell death assay detect anti-MHC α/β-Alexa 568 in channel 2 (TRITC), anti-caspase-3 Alexa 647 in channel 3 (Cy5), and BOBO-1 in channel 4 (FITC).

6. For proliferation and colony formation assays detect anti-MHC α/β-Alexa 568 in channel 2 (FITC), and Ki67 in channel 3 (TRITC).

3.5 Cell Death and Hypertrophy Assay on Live Pure hPSC-CM

Collect cell death-related nuclear events in channel 1 (Hoechst), apoptosis-related events in channel 2 (caspases 3&7), hypertrophy (Cell Tracker) or mitochondrial depolarization (TMRM) events in channel 3, and necrosis-related events in channel 4 (ToPRO-3). Use SpotDetector BioApplication algorithm (Fig. 3) (4).

1. Collect nuclear area, nuclear shape, and Hoechst intensity in channel 1. Define smaller nuclei as pyknotic nuclei (chromatin condensation), larger nuclei and extreme nuclear shape index as fragmented nuclei, lower Hoechst intensity as degraded DNA, and brighter nuclei as pyknotic/permeabilized cells.

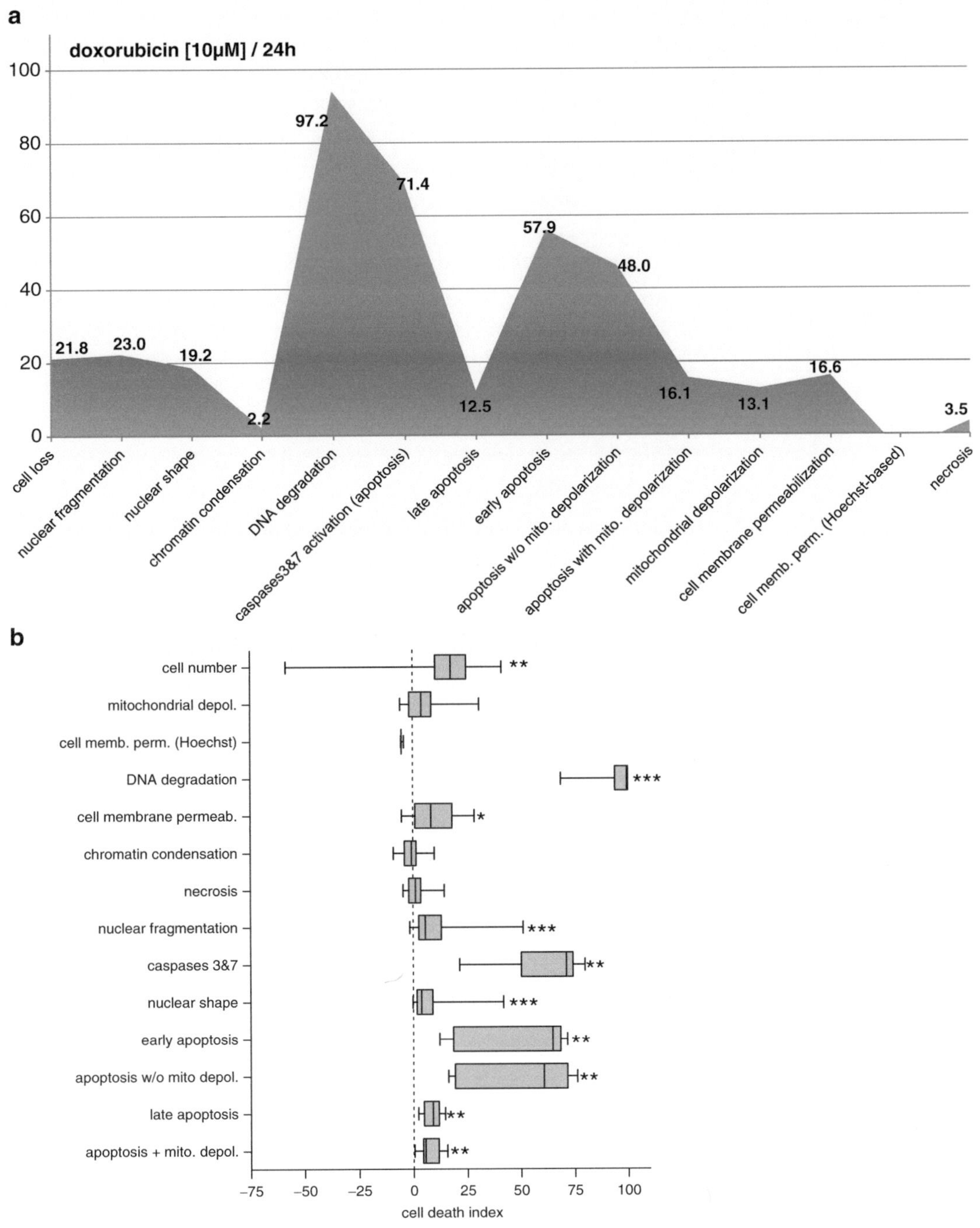

Fig. 3 (**a**) Line chart showing multivariate cell death profile of doxorubicin in hiPSC-CM. Doxorubicin induced cardiomyocyte death with clear apoptosis-specific signature consisting of DNA degradation, and early apoptosis (caspase 3&7 substrate without cell membrane permeabilization) with intact mitochondrial membrane potential. *Axis* shows cell death index, with zero representing the control and 100 the maximal effect. $n = 22$ from four experiments. (**b**) *Box plots* showing statistical significance of doxorubicin-induced cell death. Mean, upper, and lower quartiles as well as extreme values are plotted, showing the robustness of various cell death readouts. Readouts were sorted on their standard deviation in control condition (not shown), from the less reliable (cell loss) to the most reliable (apoptosis with mitochondrial depolarization)

2. Determine the intensity threshold value in channel 2 for nuclear and perinuclear high caspase intensity. In control, the majority of cells should be caspase negative. After scanning the control wells, determine the spot area threshold that gives 5 % caspase-positive cells. Collect caspase high-spot area.

3. Use the Spot Detector tool to overlap the spots with Cell Tracker staining in channel 3. Collect the percentage of cells with larger areas as hypertrophy-readout and the percentage of cells with smaller area which can characterize apoptotic cells.

4. Alternatively, if hypertrophy is not required, set TMRM spot area (mitochondrial depolarization) and collect only low TMRM area. The negative TMRM population is typically 25 % in control (see Note 7).

5. Similarly to step 4, collect spot area in channel 4 after determining the intensity/area thresholds. Measure spots in the nucleus only.

6. Generate additional functional readouts using the Event tool: Event1: early apoptosis (high caspase AND NOT high ToPRO-3); Event 2: apoptosis without mitochondrial depolarization (high caspase AND NOT low TMRM); and Event 3: necrosis (high ToPRO-3 AND NOT high caspase). Generate late apoptosis and apoptosis group data with mitochondrial depolarization by subtracting high caspase with early apoptosis and high caspase with apoptosis without mitochondrial depolarization, respectively.

3.6 Cell Death Assay on Fixed Population of hPSC-CM

Fix and label cells with anti-MHC and anti-caspase-3 antibodies for immunocytochemistry. Collect cell death-related nuclear events in channel 1 (Hoechst), cell remodelling events in channel 2 (MHC), apoptosis-related events in channel 3 (caspase-3), and necrosis-related events in channel 4 (BOBO-1). For data collection, use SpotDetector BioApplication. Further details on live and fixed cell death assays are given elsewhere (4).

1. Collect nuclear events as described under Section 3.5, step 1.

2. Match spot in channel 2 with MHC staining and collect spot sizes and intensity features. Reject object without spots (i.e., non-cardiomyocytes) from the analysis. Thus, nuclear events, apoptosis, and necrosis will only be measured in cardiomyocytes.

3. Collect caspase-3 spot high intensity/area as described under Section 3.5, step 4.

4. Collect BOBO-1 spot high intensity/area as described for ToPRO-3 in Section 3.5, step 6 (see Note 8).

5. Generate additional functional readouts using the Event tool. Event 1: early apoptosis (high caspase-3 AND NOT high

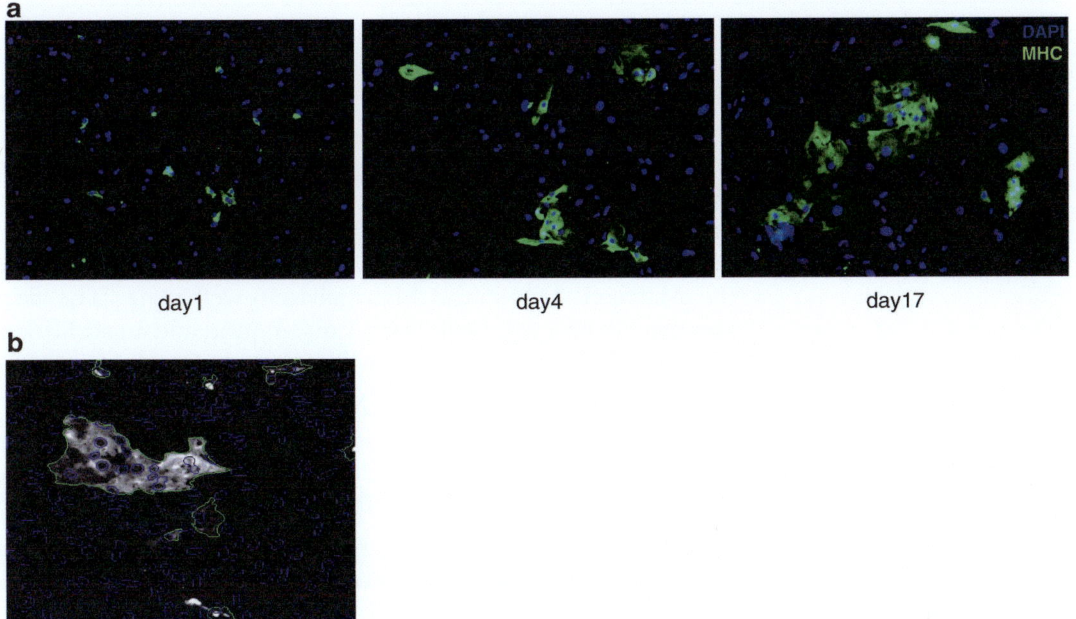

Fig. 4 (**a**) Representative images of human embryonic stem cell-derived cardiomyocytes and clusters plated (*green*, anti-MHC antibody; *blue*, Hoechst). One day following plating, cardiomyocytes are seen as individual cells; the rate of proliferation in these cells is too low to allow formation of de novo colony. At day 4, some myocytes proliferated and formed small colonies. At day 17, larger colonies are detected. (**b**) Using Colony Formation algorithm, colonies are detected as primary objects and the number of nuclei within the colony counted. The actual number of cells forming a colony can be deduced and thus the rate of proliferation of hESC-CM

BOBO-1); Event 2: cell remodelling (high and low MHC areas OR high and low MHC intensities); and Event 3: necrosis (high BOBO-1 AND NOT high caspase-3). Deduce late apoptosis by subtracting high caspase-3 with early apoptosis readout.

3.7 Cell Proliferation in Fixed Cardiomyocyte Cultures

1. Fix cells after drug treatment and stain with anti-MHC-Alexa 488, anti-Ki67-Alexa 568, and Hoechst (Fig. 4a).

2. Quantitate DNA content and visualize DNA intensity as a histogram. Quantitate percentage of 2 and 4 N DNA content subpopulations. Quantitate the ratio of Ki67-positive nuclei in the culture to assess proliferating fraction of the population.

3. Rescan the plate using Colony Formation BioApplication with MHC in channel 1 and Hoechst in channel 2. Determine one colony (anti-MHC-positive cells) as a primary object and collect the number of nuclei per colony using a spot detection tool (Fig. 4b) (see Note 9).

4 Notes

1. Since Matrigel is liquid at 4 °C, and gels rapidly at room temperature, it is critical not to permit it to warm during preparation. Place a sterile 15 ml conical tube and cold DMEM/F-12 medium into the biosafety cabinet. After removing Matrigel aliquot from −20 °C freezer add 1 ml of cold DMEM/F-12 to it. Pipette up and down to thaw and dissolve the Matrigel. Dilute the Matrigel to 6-well plate. The final volume will be 1 ml/well. For a 1 mg aliquot, add an additional 11 ml of DMEM/F-12. This will make two 6-well plates.

2. Alternatively, purified hiPSC-CM (red fluorescent protein-free) cardiomyocytes can be purchased from commercial suppliers such as Cellular Dynamics International and human embryonic stem cell-derived cardiomyocytes from GE Healthcare.

3. Doxorubicin is harmful; wear gloves and, when weighing the powder, use respiratory protection. Doxorubicin may not dissolve rapidly when diluted in RPMI supplemented with B27. If this happens, warm up the solution at 37 °C for 5–10 min.

4. Time-lapse microscopy: Observe the number of beating cells, increase in cell size, and beating activity. Use time-lapse video recordings in a live-cell imaging chamber with constant 37 °C and physiological pH. Obtain images and cell length measurements by using an inverted light microscope (Nikon TE 2000) with Elements AR software. Determine cell surface area by computer-assisted planimetry.

5. Alternatively, cardiomyocytes can be generated from embryoid body-based method (5). Medium for embryoid body-based differentiation: 400 ml of knockout DMEM 100 ml of heat-inactivated fetal bovine serum—20 %, 5 ml nonessential amino acids—1 %, 2.5 ml penicillin/streptomycin (Gibco)—0.5 %, 2.5 ml L-glutamine—0.5 %, and 35 µl β-mercaptoethanol (dilute stock 1:100 in H_2O before use).

6. In order to eliminate the remaining undissociated clusters, cell suspension can be filtered through 40 µm cell strainer (BD Biosciences).

7. TMRM is naturally extruded by cells through multidrug resistance pumps resulting in a mixed population of depolarized cells and cells that extruded the dye in the TMRM-negative population (6). Although it is not possible to distinguish two subpopulations in control, an increase in the number of TMRM-negative cells is assumed to reflect a defect in mitochondrial function.

8. Following fixation, BOBO-1 staining spreads out across cells resulting in homogenization in staining. However, in genuine necrotic cells, BOBO-1 fluorescence intensity remains higher than in non-necrotic cells.

9. As hPSC-CM are mononucleated, the number of nuclei per colony gives an accurate estimation of the number of cells per colony (7).

Acknowledgements

G.F. was supported by Fondation Leducq and Hungarian Scientific Research Fund (OTKA 105555) and TÁMOP-4.2.2/B-10/1-2010-0013 Grant.

References

1. Burridge PW, Keller G, Gold JD et al (2012) Production of de novo cardiomyocytes: human pluripotent stem cell differentiation and direct reprogramming. Cell Stem Cell 10:16–28

2. Matsa E, Denning C (2012) In vitro uses of human pluripotent stem cell-derived cardiomyocytes. J Cardiovasc Transl Res 5:581–592

3. Laflamme MA, Chen KY, Naumova AV et al (2007) Cardiomyocytes derived from human embryonic stem cells in pro-survival factors enhance function of infarcted rat hearts. Nat Biotechnol 25:1015–1024

4. Mioulane M, Foldes G, Ali NN et al (2012) Quantification of cell death mechanisms in human pluripotent stem cell-derived cardiomyocytes. J Cardiovasc Transl Res 5:593–604

5. Kehat I, Kenyagin-Karsenti D, Snir M et al (2001) Human embryonic stem cells can differentiate into myocytes with structural and functional properties of cardiomyocytes. J Clin Invest 108:407–414

6. Hattori F, Chen H, Yamashita H et al (2010) Nongenetic method for purifying stem cell-derived cardiomyocytes. Nat Methods 7:61–66

7. Foldes G, Mioulane M, Wright JS et al (2011) Modulation of human embryonic stem cell-derived cardiomyocyte growth: a testbed for studying human cardiac hypertrophy? J Mol Cell Cardiol 50:367–376

Methods Molecular Biology (2013) 1052: 41–48
DOI 10.1007/7651_2013_29
© Springer Science+Business Media New York 2013
Published online: 5 June 2013

A High Content Imaging-Based Approach for Classifying Cellular Phenotypes

Joseph J. Kim, Sebastián L. Vega, and Prabhas V. Moghe

Abstract

Current methods to characterize cell–biomaterial interactions are population-based and rely on imaging or biochemical analysis of end-point biological markers. The analysis of stem cells in cultures is further challenged by the heterogeneous nature and divergent fates of stem cells, especially in complex, engineered microenvironments. Here, we describe a high content imaging-based platform capable of identifying cell subpopulations based on cell phenotype-specific morphological descriptors. This method can be utilized to identify microenvironment-responsive morphological descriptors, which can be used to parse cells from a heterogeneous cell population based on emergent phenotypes at the single-cell level and has been successfully deployed to forecast long-term cell lineage fates and screen regenerative phenotype-prescriptive biomaterials.

Keywords Bioimage informatics, High content image analysis, Pluripotent stem cells, Phenotypic classification, Immunocytochemistry

1 Introduction

Engineered microenvironments and biomaterials are capable of controlling a wide range of cellular phenomena including attachment, cell morphology, proliferation, and lineage commitment (1–5). Characterization methods of biomaterial properties are well established, but methods to quantitatively characterize a cell's response to biomaterials are not. Traditionally, end-point biological assays have been used to determine the levels of lineage-specific markers of differentiated stem cells. These methods are time consuming and fail to elucidate the role of the microenvironment on stem cell function at the single-cell level. To accelerate biomaterial screening, methods that can simultaneously test a large number of conditions have been developed (6–8). However, these techniques continue to rely on biological assays that measure markers specific to primarily mature phenotypes as opposed to emergent phenotypes. Additionally, these methods measure the average expression of cell populations, which produce only approximate and frequently inaccurate readouts in heterogeneous cell cultures (9).

In order to address these limitations, we demonstrate a high content imaging-based platform capable of parsing numerical morphological descriptors from images of individual stem cells cultured on various microenvironments (10–12). This method is based on the notion that there are several key cytoskeletal and nuclear proteins whose organization is sensitively influenced by a signaling pathways emanating from extracellular and environmental stimuli. The expression of these proteins can be captured via fluorescent labeling and high content imaging, and ultimately quantified by morphological descriptors, which provide a "barcode-like" identity to single cells. By employing this approach, insights about the organization of various cellular proteins can be realized at a single-cell level, and contextualized and parsed around the nature of the cellular environment, and tracked as a function of cellular development kinetics.

To date, the high content imaging-based approach has been applied to a number of cell systems (e.g., mesenchymal stem cells (10–12), embryonic stem cells (10), induced pluripotent stem cells (10), and breast epithelial cancer cells (13)) cultured in different microenvironments (e.g., orthopedic implants (14), PEG-variant biomaterials (10), and 3D Matrigel (13)), illustrating its versatility in biological research. The workflow procedure of this approach is very straightforward and relatively inexpensive (Fig. 1), which allows for virtually any laboratory equipped with a fluorescent microscope to be able to perform this technique. More sophisticated microscopes (e.g., confocal laser scanning and super-resolution instruments) are able to achieve images of higher spatial resolution, which subsequently provide greater detail required for sub-cellular structures, but this is not necessary for larger organelles or protein distributions in the cell body and in the nuclear space.

2 Materials

Virtually any adherent cell type that can be cultured in vitro and immunocytochemically labeled can be used for this assay. As an example, we will provide the materials required for imaging of cultured human mesenchymal stem cells derived from the bone marrow. Unless indicated otherwise, store all reagents at 4 °C. Follow appropriate waste disposal protocols with all materials after their use.

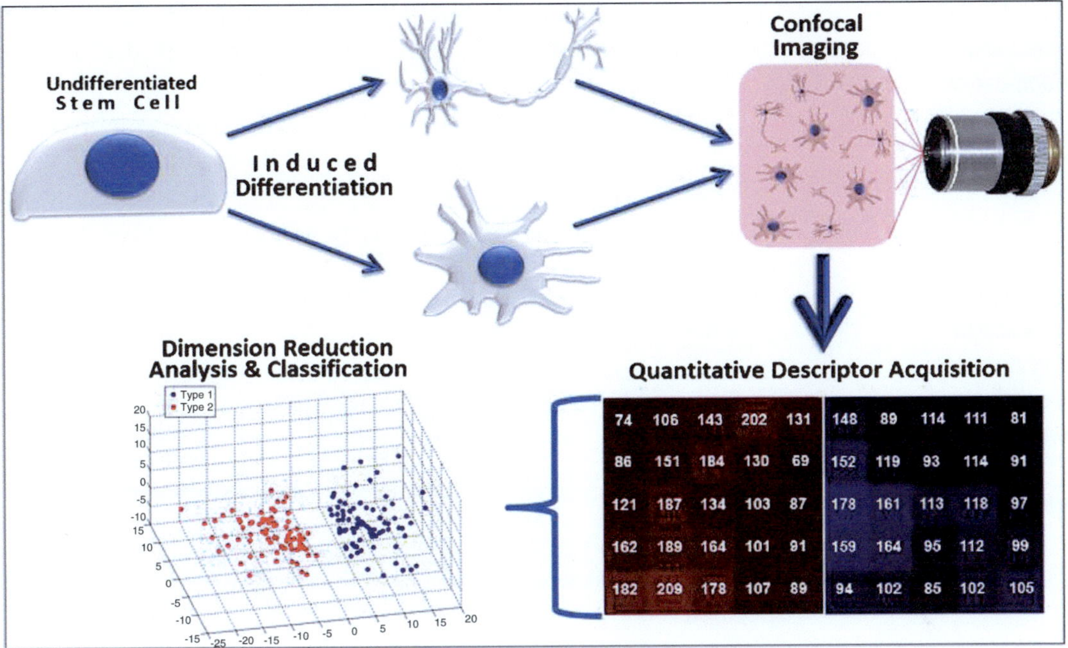

Fig. 1 Workflow schematic for high content imaging-based characterization of different cellular phenotypes

2.1 Mesenchymal Stem Cell Culture Medium Formulations

1. Basal Mesenchymal Stem Cell Maintenance Medium: Minimum Essential Medium (MEM)-α without deoxy- or ribonucleosides (Gibco, CA), supplemented with 10 % fetal bovine serum (FBS) (Gibco, CA) and 100 units/mL penicillin plus 100 μg/mL streptomycin (Gibco, CA).

2. Osteogenesis Induction Medium: MEM-α without deoxy- or ribonucleosides, supplemented with 10 % FBS, 10 mM β-glycerophosphate (Sigma-Aldrich, Saint Louis, MO), 0.1 μM dexamethasone (Sigma-Aldrich, Saint Louis, MO), 0.5 mM L-ascorbic acid-2-phosphate (Sigma-Aldrich, Saint Louis, MO), and 100 units/mL penicillin plus 100 μg/mL streptomycin.

3. Adipogenesis Induction Medium: MEM-α without deoxy- or ribonucleosides, supplemented with 10 % FBS, 1 μM dexamethasone, 0.2 mM indomethacin (Sigma-Aldrich, Saint Louis, MO), 0.01 mg/mL insulin (Sigma-Aldrich, Saint Louis, MO), 0.5 mM 3-isobutyl-1-methyl-xanthine (Sigma-Aldrich, Saint Louis, MO), and 100 units/mL penicillin plus 100 μg/mL streptomycin.

4. Adipogenic Maintenance Medium: MEM-α without deoxy- or ribonucleosides, supplemented with 10 % FBS, 0.01 mg/mL insulin (Sigma-Aldrich, Saint Louis, MO) and 100 units/mL penicillin plus 100 μg/mL streptomycin.

<table>
<tr><td>

2.2 Immunocyto-
chemistry
Components

</td><td>

1. Permeabilization Buffer: Phosphate Buffered Saline (PBS) (Lonza) without Ca^{2+}, Mg^{2+} supplemented with 0.1 % Triton X-100.

2. Blocking Buffer: PBS without Ca^{2+}, Mg^{2+} supplemented with 5 % Normal Goat Serum (NGS) (MP Biomedicals).

3. Primary and Secondary Antibody Solutions: Dilute antibodies in blocking buffer to antibody-specific concentrations (see Note 1).

</td></tr>
<tr><td>

2.3 Image
Acquisition
and Analysis
Workstation
Components

</td><td>

1. Microscope equipped with multiple fluorescence filters and/or photodetectors.

2. Computer equipped with Image Processing software (ImageJ was selected as the illustrative example).

</td></tr>
</table>

3   Methods

Carry out all procedures at room temperature unless otherwise specified.

<table>
<tr><td>

3.1 Cell Culture

</td><td>

1. Culture hMSCs until they reach 50 % confluency (see Note 2).

2. Remove medium and wash cells 1× with pre-warmed PBS (Lonza).

3. Remove PBS and add trypsin (Sigma) to incubate for 5 min.

4. Gently tap expansion flask and verify cell detachment under a light microscope.

5. Add equal volume of basal mesenchymal stem cell maintenance medium to neutralize trypsin.

6. Transfer contents of flask to a 15 mL conical tube and centrifuge at 600 RCF for 5 min.

7. Resuspend in 1 mL of basal mesenchymal stem cell maintenance medium.

8. Determine viable cell count via Trypan Blue Stain (Lonza) on a hemacytometer (see Note 3).

9. Seed cells (density = 5,000 cells/cm^2) onto a glass surface (e.g., glass-bottom plate).

10. Allow cells to attach for 6 h and begin differentiation induction.

</td></tr>
<tr><td>

3.2 Immunocyto-
chemistry and Imaging

</td><td>

1. Once cells have attached and reached the desired phenotype, wash cells with PBS and fix cells with 4 % paraformaldehyde (Electron Microscopy Sciences) for 15 min.

2. Wash cells 3× with PBS for 5 min per wash.

3. Block and permeabilize samples with 0.1 % Triton X-100 (Sigma) + 5 % NGS (MP Biomedicals) in PBS for 1 h.

</td></tr>
</table>

4. After two washes with blocking buffer (5 % NGS in PBS), add primary antibody solution at optimal concentration and incubate overnight at 4 °C with agitation (see Note 4).

5. Remove primary antibody solutions and perform three 15 min washes in blocking buffer.

6. Add secondary fluorophore antibody solution at optimal concentration for 2 h at room temperature with agitation (see Note 5).

7. Remove secondary antibody solutions and wash with PBS for 2 h, changing PBS every 30 min.

8. Add 1 μg/mL DAPI (Sigma) in PBS for 5 min.

9. Wash cells 2× with PBS.

10. Leave cells in PBS and cover in foil. Store at 4 °C until ready for imaging.

11. Utilize an epifluorescence or confocal fluorescence microscope. Acquire images with a high-magnification objective lens (at least 40×) using the appropriate fluorescence filter cubes or laser excitation wavelengths that correspond to your secondary antibody fluorophores. Save images on a computer or portable hard drive such that these can be recalled for further image processing.

3.3 Quantitative Descriptor Acquisition Using an Image Processing Software (This Example Is Based on ImageJ)

1. Obtain the latest version of software. For example, ImageJ was used as the software of choice. This can be downloaded and installed from the Research Services Branch of the National Institute of Health in the US (http://rsb.info.nih.gov/ij/).

2. Make sure your image(s) is in a format recognizable by ImageJ and open it (File:Open).

3. Set the upper and lower threshold according to your label of interest in the "Threshold" window (Image:Adjust:Threshold), and click "Apply." This will convert any pseudo-colored image to grayscale (see Fig. 2).

4. Choose quantitative descriptors of interest in the "Set Measurements" window (Analyze:Set Measurements).

5. Acquire measurements by accessing the "Analyze Particles" window (Analyze:Analyze Particles) and check the following options: Display results, Clear Results & Summarize. Leave every other option unchecked and click "OK."

6. Selected quantitative descriptors will be measured and displayed in the "Results" window (Fig. 3). This data can be saved as a spreadsheet (File:Save As). ImageJ offers several options to analyze this data in the "Results" tab of the "Results" window.

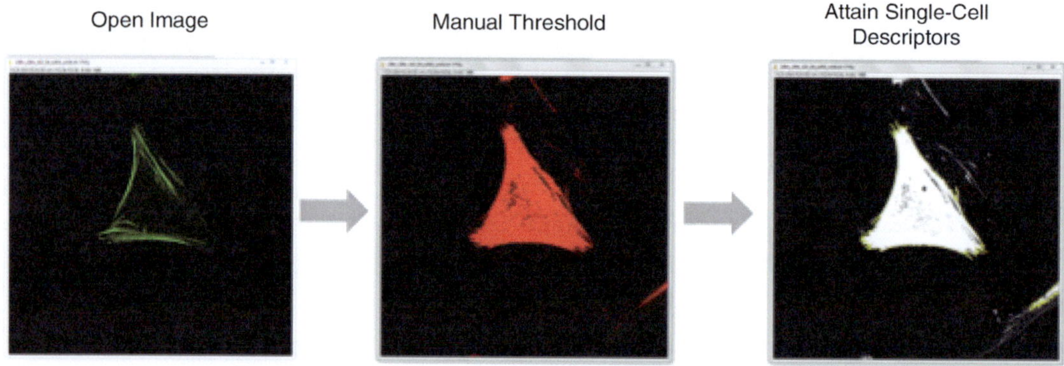

Fig. 2 Basic digital image processing steps on high-resolution micrographs of cells yield images of single cells and intracellular organization that can be subsequently processed to yield cell descriptors

3.4 Dimension Reduction Analysis Using MATLAB

1. Open Matlab (Mathworks) equipped with their "Statistics Toolbox."

2. Input your quantitative descriptor data banks as matrices, with each column corresponding to a different descriptor and each row corresponding to a different single cell. Assign different variables corresponding to different cell phenotypes or conditions.

3. Obtain the Matlab Toolbox for Dimensionality Reduction (http://www.mathworks.com/matlabcentral/linkexchange/links/1626-the-matlab-toolbox-for-dimensionality-reduction OR http://cseweb.ucsd.edu/~lvdmaaten/dr/download.php)

4. Implement dimension reduction algorithms via multidimensional scaling or principal component analysis (see Note 6; Fig. 3).

4   Notes

1. For best results for immunocytochemistry, try incubation with several concentrations of antibody based on recommended levels for similar protein(s) and cell type(s) of interest and maximize signal-to-noise ratio relative to background staining and negative controls (isotype and secondary alone controls).

2. Culture hMSCs in expansion flasks (e.g., T25s, T75s). High content imaging is best suited for sub-confluent seeding densities. 50 % confluency is generally achieved when in any given field of view 50 % is cells, and 50 % is void space.

3. First aliquot a small, yet representative sample of cell suspension. Next, add Trypan Blue Stain and note the dilution factor. Then, add 10 µL to a standard hemocytometer. To calculate

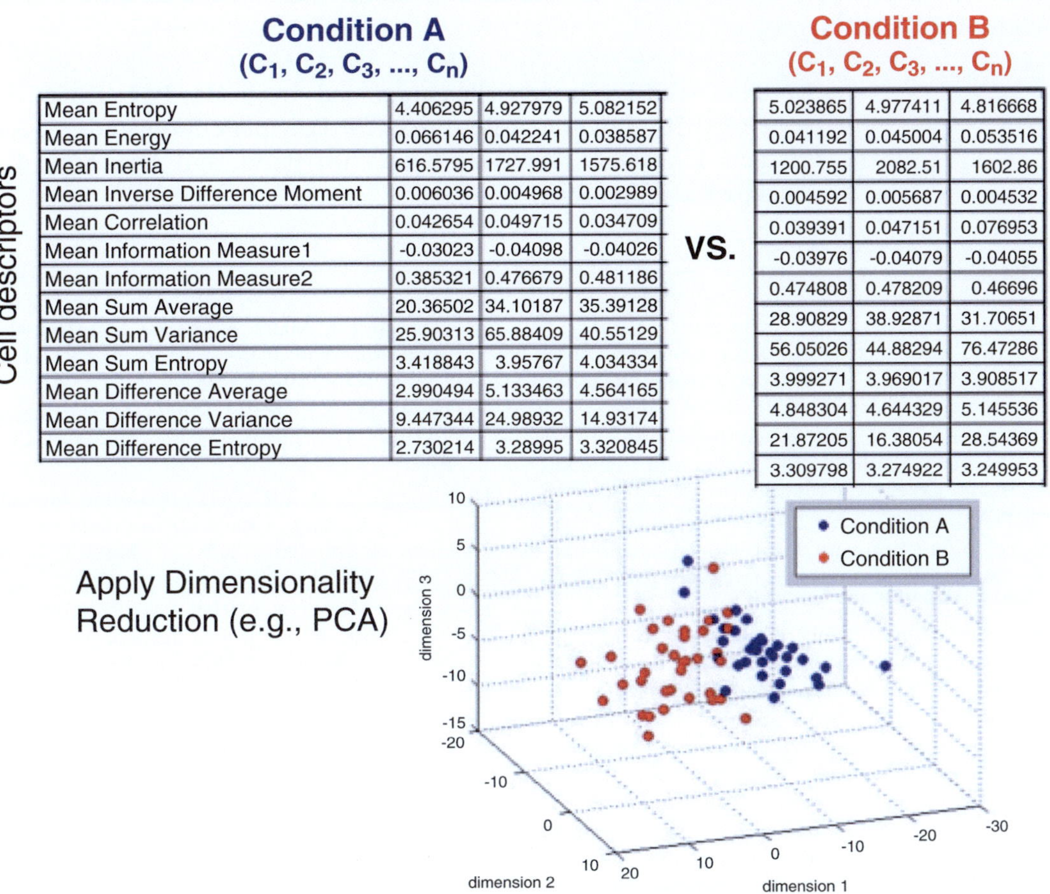

Condition A (C_1, C_2, C_3, ..., C_n)

Mean Entropy	4.406295	4.927979	5.082152
Mean Energy	0.066146	0.042241	0.038587
Mean Inertia	616.5795	1727.991	1575.618
Mean Inverse Difference Moment	0.006036	0.004968	0.002989
Mean Correlation	0.042654	0.049715	0.034709
Mean Information Measure1	-0.03023	-0.04098	-0.04026
Mean Information Measure2	0.385321	0.476679	0.481186
Mean Sum Average	20.36502	34.10187	35.39128
Mean Sum Variance	25.90313	65.88409	40.55129
Mean Sum Entropy	3.418843	3.95767	4.034334
Mean Difference Average	2.990494	5.133463	4.564165
Mean Difference Variance	9.447344	24.98932	14.93174
Mean Difference Entropy	2.730214	3.28995	3.320845

Condition B (C_1, C_2, C_3, ..., C_n)

5.023865	4.977411	4.816668
0.041192	0.045004	0.053516
1200.755	2082.51	1602.86
0.004592	0.005687	0.004532
0.039391	0.047151	0.076953
-0.03976	-0.04079	-0.04055
0.474808	0.478209	0.46696
28.90829	38.92871	31.70651
56.05026	44.88294	76.47286
3.999271	3.969017	3.908517
4.848304	4.644329	5.145536
21.87205	16.38054	28.54369
3.309798	3.274922	3.249953

Fig. 3 Single-cell descriptors can be obtained and compared between two different treatment conditions following dimensionality reduction techniques such as principal component analysis (PCA). The three dimensions on the graph for example represent the three most influential combinations of cell descriptors that can parse the two cell populations shown in *blue* and *red*

cell solution concentration (number of cells per mL): (a) divide cell count by number of quadrants counted, (b) multiply by dilution factor, and (c) multiply by 10,000.

4. It is recommended to use a standard rocker for this step.

5. This step and all subsequent steps should be performed with minimal light exposure in order to prevent bleaching of the secondary antibody fluorophore. At the final step, use an anti-fade reagent to preserve samples for prolonged or repeated imaging.

6. The output values of sensitivity and specificity provide statistical measures of classification accuracy.

Acknowledgments

This study was partially supported by NIH P41 EB001046 (RESBIO, Integrated Resources for Polymeric Biomaterials), Rutgers University Academic Excellence Fund, and NSF Stem Cell IGERT 0801620.

References

1. Causa F, Netti PA, Ambrosio L (2007) A multi-functional scaffold for tissue regeneration: the need to engineer a tissue analogue. Biomaterials 28(34):5093–5099

2. Guarino V, Causa F, Ambrosio L (2007) Bioactive scaffolds for bone and ligament tissue. Expert Rev Med Devices 4(3):405–418

3. Pasquinelli G et al (2008) Mesenchymal stem cell interaction with a non-woven hyaluronan-based scaffold suitable for tissue repair. J Anat 213(5):520–530

4. Reed CR et al (2009) Composite tissue engineering on polycaprolactone nanofiber scaffolds. Ann Plast Surg 62(5):505–512

5. Shea LD et al (2000) Engineered bone development from a pre-osteoblast cell line on three-dimensional scaffolds. Tissue Eng 6(6):605–617

6. Weber N et al (2004) Small changes in the polymer structure influence the adsorption behavior of fibrinogen on polymer surfaces: validation of a new rapid screening technique. J Biomed Mater Res 68(3):496–503

7. Flaim CJ, Chien S, Bhatia SN (2005) An extracellular matrix microarray for probing cellular differentiation. Nat Methods 2(2):119–125

8. Dittrich PS, Manz A (2006) Lab-on-a-chip: microfluidics in drug discovery. Nat Rev Drug Discov 5(3):210–218

9. Levsky JM, Singer RH (2003) Gene expression and the myth of the average cell. Trends Cell Biol 13(1):4–6

10. Vega SL et al (2012) High-content imaging-based screening of microenvironment-induced changes to stem cells. J Biomol Screen 17(9):1151–1162

11. Treiser MD et al (2010) Cytoskeleton-based forecasting of stem cell lineage fates. Proc Natl Acad Sci U S A 107(2):610–615

12. Liu E et al (2010) Parsing the early cytoskeletal and nuclear organizational cues that demarcate stem cell lineages. Cell Cycle 9(11):2108–2117

13. Vidi PA et al (2012) Interconnected contribution of tissue morphogenesis and the nuclear protein NuMA to the DNA damage response. J Cell Sci 125(Pt 2):350–361

14. McNamara LE et al (2011) Skeletal stem cell physiology on functionally distinct titania nanotopographies. Biomaterials 32(30):7403–7410

Methods Molecular Biology (2013) 1052: 49–56
DOI 10.1007/7651_2013_24
© Springer Science+Business Media New York 2013
Published online: 3 May 2013

Conversion of Primordial Germ Cells to Pluripotent Stem Cells: Methods for Cell Tracking and Culture Conditions

Go Nagamatsu and Toshio Suda

Abstract

Primordial germ cells (PGCs) are unipotent cells committed to germ lineage: PGCs can only differentiate into gametes in vivo. However, upon fertilization, germ cells acquire the capacity to differentiate into all cell types in the body, including germ cells. Therefore, germ cells are thought to have the potential for pluripotency. PGCs can convert to pluripotent stem cells in vitro when cultured under specific conditions that include bFGF, LIF, and the membrane-bound form of SCF (mSCF). Here, the culture conditions which efficiently convert PGCs to pluripotent embryonic germ (EG) cells are described, as well as methods used for identifying pluripotent candidate cells during culture.

Keywords: Primordial germ cell, Acquisition of pluripotency, Tracking, Purification

1 Introduction

Germ cells are committed cells that have a unipotent differentiation capacity, giving rise only to gametes even though they have the developmental competence to form cells of all three germ layers upon fertilization (1, 2). Therefore, germ cells are thought to have pluripotent potential. Primordial germ cells (PGCs) express key factors that maintain pluripotency, such as Oct3/4, Sox2, and Nanog (3–5). Furthermore, PGCs can be converted to pluripotent embryonic germ (EG) cells under appropriate culture conditions (6). Thus, the conversion of PGCs to EG cells represents the process of pluripotency acquisition by the unipotent germ cell lineage. This PGC to EG conversion will provide a good model for analyzing how pluripotency is acquired, and how PGCs control the balance between pluripotency and unipotency. Intriguingly, the functional

The online version of this chapter (doi: 10.1007/7651_2013_24) contains supplementary material, which is available to authorized users.

genes involved in germ cell development also have reprogramming activity in somatic cells (7). Thus, understanding the process of germ cell conversion to pluripotency could be useful for somatic cell reprogramming.

To understand the mechanisms underlying the acquisition of pluripotency, it is critical to identify cells that can be converted to pluripotent stem cells. First, efficient culture conditions that induce pluripotency from PGCs were established (8). By screening a number of chemical compounds, including epigenetic modifiers such as trichostatin A (TSA) and 5-azacytidine (5AZA), we identified three inhibitors that were highly efficient in stimulating the conversion of PGCs to EG cells when added to the culture medium. In these cultures, the generation of EG cells from PGCs was ~15 % efficient. The three inhibitors are the MEK inhibitor, PD325901, the GSK-3β inhibitor, CHIR99021, and the TGF-βR inhibitor, A83-01. The combination of the MEK and GSK-3β inhibitors has been referred to as 2i in pluripotent stem cell studies, and our three-inhibitor combination is designated 2i + A83 (9, 10).

Fluorescence-activated cell sorting (FACS) is used to monitor the conversion of candidate cells to a pluripotent state (8). Nanog-GFP+ and SSEA-1+ cells (identified by FACS) are collected at day 3 and day 6 of culture and replated in secondary cultures. Only this double-positive population generates EG cells, indicating that this population is in the process of converting to a pluripotent state. The high efficiency of conversion to pluripotency, and the purification of Nanog-GFP+/SSEA-1+ cells in this system, will provide novel insight into how pluripotency is acquired by germline-committed cells.

2   Materials

1. Knockout DMEM (Invitrogen; catalogue no. A1286101).

2. Knockout Serum Replacement (KSR) (Invitrogen; catalogue no. 10828–028). Make 75 ml aliquots and store at −30 °C.

3. L-Glutamine, 200 mM (Invitrogen; catalogue no. 25030–081). Make 5 ml aliquots and store at −30 °C.

4. MEM non-essential amino acids (NEAA) (Invitrogen; catalogue no. 11140–050). Store at 4 °C.

5. β-Mercaptoethanol, 55 mM (Invitrogen; catalogue no. 21985–023). Store at 4 °C.

6. ESGRO, 10^7 U/ml (Millipore; ESG1107). Make 50 µl aliquots and store at 4 °C.

7. bFGF (Peprotech; catalogue no. 450–33). Prepare a 1.0 mg/ml stock solution in water. Store at −30 °C.

8. PD0325901 (JS Research Chemicals Trading; catalogue no. MEK-PD325). Prepare a 1 mM stock solution in dimethyl sulfoxide (DMSO). Store at −30 °C.

9. CHIR99021 (Biovision; catalogue no. 1667-5). Prepare a 3 mM stock solution in DMSO. Store at −30 °C.

10. A83-01 (Stemgent, catalogue no. 130-095-565). Prepare a 250 µM stock solution in DMSO. Store at −30 °C.

11. 0.05 % Trypsin–EDTA (Invitrogen; catalogue no. 25300–054).

12. DMEM (Invitrogen; catalogue no. 10313–039). Store at 4 °C.

13. Mitomycin C, 10 mg (Kyowa Hakko Kirin). Prepare a 400 µg/ml stock solution in water. Store at −30 °C.

14. Penicillin/Streptomycin (Meiji). Make 1 and 5 ml aliquots and store at −80 °C (see Note 1).

15. Gelatin from porcine skin type A (Sigma; catalogue no. G1890). Prepare a 0.1 % solution in water. Autoclave. Store at room temperature.

16. Propidium iodide, 1.0 mg/ml (Invitrogen; catalogue no. P3586). Store at 4 °C.

17. 24-well multi-well plates (BD Falcon; catalogue no. 353047).

18. 50 ml conical centrifuge tubes (BD Falcon; catalogue no. 352070).

19. Cell strainers (BD Falcon; catalogue no. 352340).

20. 5 ml polystyrene round-bottom tubes with cell strainer caps (BD Falcon; catalogue no. 352235).

21. Anti-FcγR antibody (2.4 G2) (BD Bioscience, catalogue no. 553142).

22. Alexa Fluor 647-conjugated anti-human/mouse SSEA1 (BioLegend; catalogue no. 125608).

2.1 Culture Medium

1. ES medium and EG induction medium:
The ES medium comprises knockout DMEM containing 15 % KSR, 2 mM glutamine, 1 mM NEAA, 5.5 µM β-mercaptoethanol, penicillin/streptomycin, and 10^4 U/ml LIF. EG induction medium is made by adding 20 ng/ml bFGF to the ES medium. To prepare 2i + A83 medium, 1 µM PD0325901, 3 µM CHIR99021, and 250 nM A83-01 are added to the EG induction medium. Store at 4 °C.

2. Feeder medium:
DMEM containing 10 % FCS and penicillin/streptomycin. Store at 4 °C.

3. FACS buffer:
PBS containing 2 % FCS.

2.2 Mice	Nanog-GFP-IRES-puro transgenic mice (RBRC02290) can be obtained from RIKEN BRC, which participates in the National Bio-Resource Project of the Ministry of Education, Culture, Sports, Science and Technology (MEXT), Japan. PGCs were prepared by crossing Nanog-GFP male mice with ICR female mice. The day on which females presented a vaginal plug was considered embryonic day 0.5 (E0.5).

3   Methods

This protocol is summarized in Fig. 1.

3.1 Preparation of Feeder Cells (see Note 2)

Sl/Sl^4-m220 cells are treated with 5 μg/ml mitomycin C for 1 h and plated at a density of 4×10^5 cells/well in 24-well plates 1 day before use. STO cells are treated with 12 μg/ml mitomycin C for 2 h and plated at a density of 1×10^6 cells/55 cm^2.

3.2 Isolation of PGCs

This protocol uses Nanog-GFP reporter mice (in which endogenous Nanog expression is monitored by GFP fluorescence), as well as wild-type (WT) mice. Although both types of mice are useful for

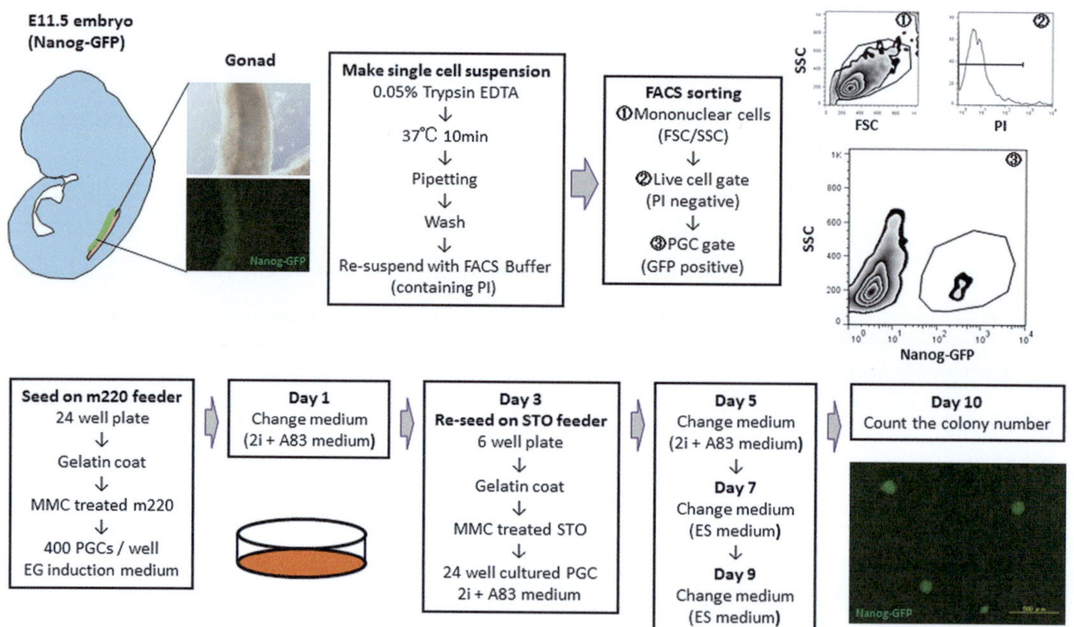

Fig. 1 Summary of the procedure used to generate EG cells from PGCs. The gonads are surgically isolated from E11.5 Nanog-GFP transgenic embryos. After preparing a single-cell suspension, the PGCs are isolated using a FACSAriaII cell sorter based on Nanog-GFP fluorescence. Purified PGCs are seeded onto m220 feeder cells. After passage of the cultured PGCs to STO feeder cells, EG cell colonies can be counted on day 10 of culture. 2i (MEK inhibitor (PD325901) and GSK-3β inhibitor (CHIR99021)). A83 (TGF-βR inhibitor (A83-01)). MMC (Mitomycin C)

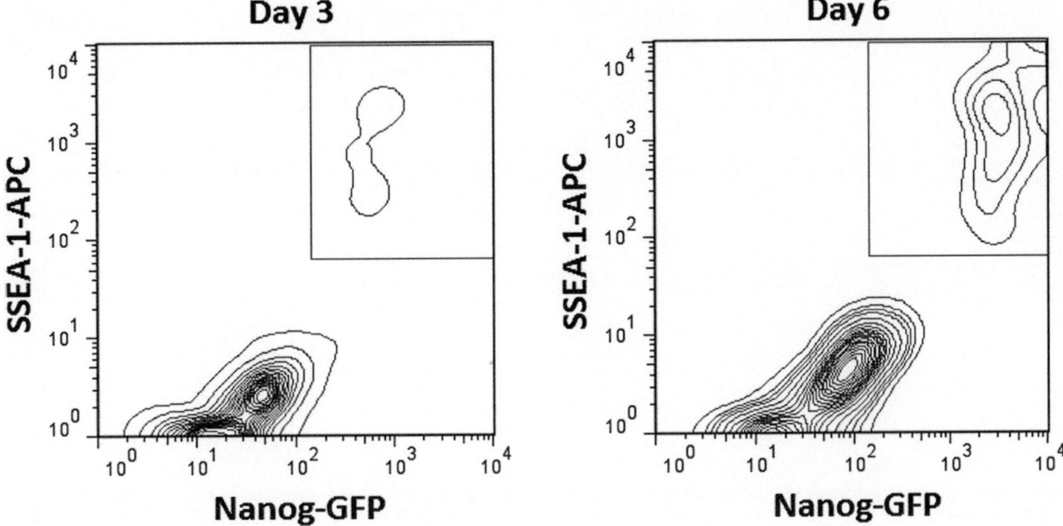

Fig. 2 Tracking pluripotent candidate cells. Shown are Nanog-GFP-positive and SSEA1-positive cells detected by FACS on day 3 and day 6 of culture to induce EG cell conversion from PGCs. Only double-positive cells (*upper right quadrant*) can form EG cells

this method, PGCs can be purified from the GFP reporter mice based on their fluorescence. When WT mice are used, the cells need to be stained for SSEA-1.

Gonads from Nanog-GFP transgenic embryos (E11.5) are dissociated to form a single-cell suspension by incubation with 0.05 % trypsin for 10 min at 37 °C (see Note 3). After washing, the cells are resuspended with FACS buffer containing 1.0 μg/ml propidium iodide (PI). PI-positive cells are dead. Therefore, PGCs are sorted as PI-negative/Nanog-GFP-positive cells using a BD FACSAriaII cell sorter (BD Bioscience). For WT embryos, PGCs are first stained for SSEA-1 as described below (see Section 3.4).

3.3   EG Cell Induction

Sorted PGCs are cultured in EG induction medium on an Sl/Sl^4-m220 feeder layer in a 24-well plate at a density of 400 PGCs per well. The next day, the medium is changed to 2i + A83 medium. On day 3, the cultured PGCs are collected and reseeded onto STO feeder cells in 2i + A83 medium in one well of a 6-well plate (see Note 4). The medium is changed every 2 days. On day 7, the medium is changed to ES medium. EG cell colonies can be counted at around day 10.

3.4   Isolation of Pluripotent Candidate Cells

Cultured PGCs are harvested in 0.05 % trypsin on day 3 or day 6 of culture (see Note 5) (see Fig. 2). After washing, the cells are incubated with an anti-FcγR antibody (2.4G2) (BD Bioscience) at 4 °C for 30 min. The cells are then incubated with an Alexa Fluor 647-conjugated anti-SSEA-1 mAb (MC-480) (BioLegend) for 30 min at 4 °C. The antibody is used at 0.2 μg/1 × 10^6 cells.

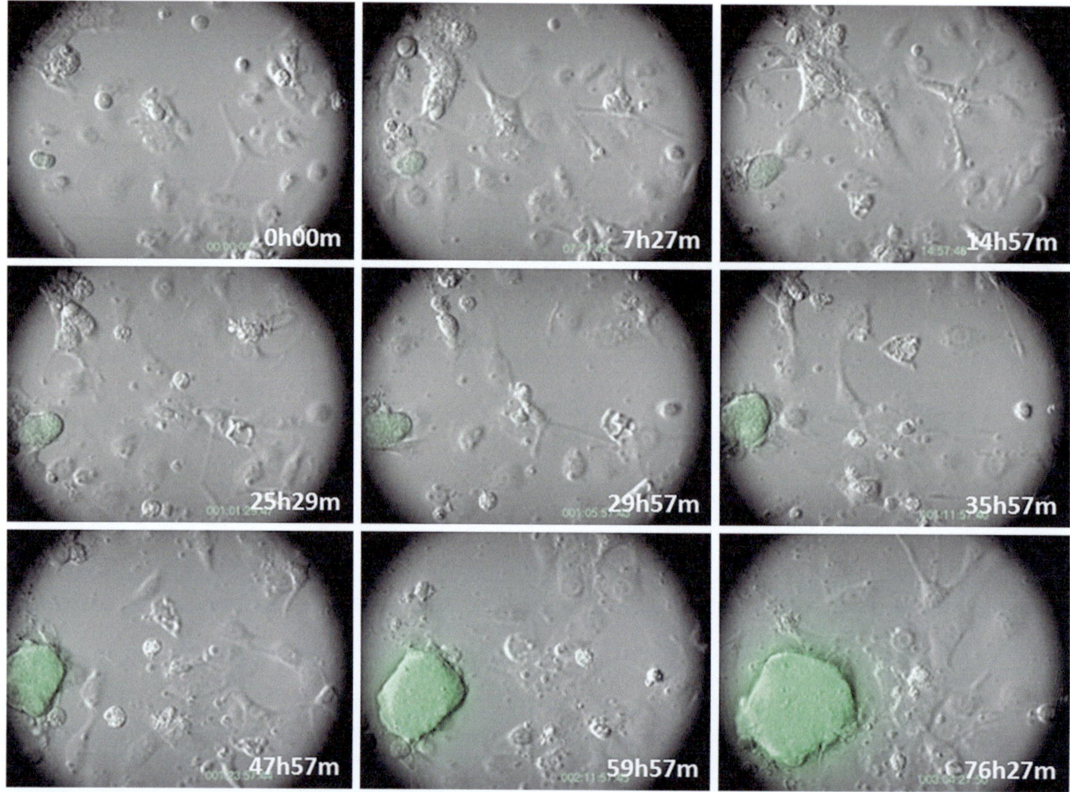

Fig. 3 Time-lapse analysis of EG cell formation. Pluripotent candidate cells, sorted on day 3 from PGCs cultured in 2i + A83 medium, were tracked by time-lapse imaging. Merged phase contrast and Nanog-GFP fluorescence images are shown. Time-lapse movie (Movie 1) is available on the Springer Extras Web site

After washing, the cells are resuspended in FACS buffer containing 1.0 µg/ml PI. The samples are analyzed and sorted using a FACSAria II cell sorter (BD Bioscience).

3.5 Time-Lapse Bio-imaging

The cultured PGCs are harvested on day 3 or 6 and reseeded onto STO feeder cells (see Fig. 3). Cells are then analyzed using an LCV110 incubator microscope system (Olympus). During the experiment, the cells are incubated at 37 °C in 5 % CO_2 (see Note 6). The collected images are analyzed using MetaMorph software (Universal Imaging, Media, PA). Time-lapse movie (Movie 1) is available on the Springer Extras Web site.

4 Notes

1. 10^6 U Penicillin (Meiji) and 1 g Streptomycin (Meiji) are dissolved in 40 ml water. Aliquots of 1 ml are stored at −30 °C. One 1-ml aliquot of the penicillin and streptomycin solution is used for every 500 ml of medium, giving final concentrations of 50 U/ml and 500 µg/ml, respectively.

2. The plates are coated with gelatin before seeding the feeder cells. To coat the culture dish, pour 0.5 or 1.0 ml of 0.1 % gelatin solution into one well of a 24-well or a 6-well culture dish, respectively, and incubate for at least 1 h at room temperature.

3. To make a single-cell suspension, gonads are washed twice with PBS and then incubated in a 0.05 % trypsin solution. After a 10-min incubation, the gonads are pipetted against the wall of the dish. Feeder medium is added to stop the reaction. The suspensions are collected in 5 ml polystyrene round-bottom tubes with a cell strainer cap.

4. PGCs cultured on the Sl/Sl^4-m220 feeder layer are removed from one well of a 24-well plate and seeded into one well of a 6-well plate containing STO feeder cells.

5. Cultured PGCs are dissociated using a 0.05 % trypsin solution. The reaction is stopped by adding feeder medium. The cell suspension is collected in a 50 ml conical centrifuge tube through a cell strainer.

6. To avoid changes in the focal plane, the culture medium is not changed during the time-lapse recording experiments.

Acknowledgements

We thank Dr. Takeo Kosaka for contributing to the establishment of this culture system. This study was supported, in part, by a grant-in-aid from the Ministry of Education, Culture, Sports, Science, and Technology of Japan; by PRESTO; and by the Takeda Science Foundation.

Electronic Supplementary Material

Below is the link to the electronic supplementary material.
Movie 1 Time-laps bio-imaging of EG cell formation. (AVI 71045 kb).

References

1. Sasaki H, Matsui Y (2008) Epigenetic events in mammalian germ-cell development: reprogramming and beyond. Nat Rev Genet 9:129–140

2. Saitou M, Kagiwada S, Kurimoto K (2012) Epigenetic reprogramming in mouse pre-implantation development and primordial germ cells. Development 139:15–31

3. Yoshimizu T, Sugiyama N, De Felice M, Yeom YI, Ohbo K, Masuko K et al (1999) Germline-specific expression of the Oct-4/green fluorescent protein (GFP) transgene in mice. Dev Growth Differ 41:675–684

4. Yamaji M, Seki Y, Kurimoto K, Yabuta Y, Yuasa M, Shigeta M et al (2008) Critical function of Prdm14 for the establishment of the germ cell lineage in mice. Nat Genet 40:1016–1022

5. Okita K, Ichisaka T, Yamanaka S (2007) Generation of germline-competent induced pluripotent stem cells. Nature 448:313–317

6. Matsui Y, Zsebo K, Hogan BL (1992) Derivation of pluripotential embryonic stem cells from murine primordial germ cells in culture. Cell 70:841–847

7. Nagamatsu G, Kosaka T, Kawasumi M, Kinoshita T, Takubo K, Akiyama H et al (2011) A germ cell-specific gene, Prmt5, works in somatic cell reprogramming. J Biol Chem 286:10641–10648

8. Nagamatsu G, Kosaka T, Saito S, Takubo K, Akiyama H, Sudo T et al (2012) Tracing the conversion process from primordial germ cells to pluripotent stem cells in mice. Biol Reprod 86(182):1–11

9. Ying QL, Wray J, Nichols J, Batlle-Morera L, Doble B, Woodgett J et al (2008) The ground state of embryonic stem cell self-renewal. Nature 453:519–523

10. Silva J, Barrandon O, Nichols J, Kawaguchi J, Theunissen TW, Smith A (2008) Promotion of reprogramming to ground state pluripotency by signal inhibition. PLoS Biol 6:e253. 10.1371/journal.pbio.0060253

Methods Molecular Biology (2013) 1052: 57–76
DOI 10.1007/7651_2013_28
© Springer Science+Business Media New York 2013
Published online: 5 June 2013

Imaging and Tracking of Bone Marrow-Derived Immune and Stem Cells

Youbo Zhao, Andrew J. Bower, Benedikt W. Graf, Marni D. Boppart, and Stephen A. Boppart

Abstract

Bone marrow (BM)-derived stem and immune cells play critical roles in maintaining the health, regeneration, and repair of many tissues. Given their important functions in tissue regeneration and therapy, tracking the dynamic behaviors of BM-derived cells has been a long-standing research goal of both biologists and engineers. Because of the complex cellular-level processes involved, real-time imaging technologies that have sufficient spatial and temporal resolution to visualize them are needed. In addition, in order to track cellular dynamics, special attention is needed to account for changes in the microenvironment where the cells reside, for example, tissue contraction, stretching, development, etc. In this chapter, we introduce methods for real-time imaging and longitudinal tracking of BM-derived immune and stem cells in in vivo three-dimensional (3-D) tissue environments with an integrated optical microscope. The integrated microscope combines multiple imaging functions derived from optical coherence tomography (OCT) and multiphoton microscopy (MPM), including optical coherence microscopy (OCM), micro-vasculature imaging, two-photon excited fluorescence (TPEF), and second harmonic generation (SHG) microscopy. Short- and long-term tracking of the dynamic behavior of BM-derived cells involved in cutaneous wound healing and skin grafting in green fluorescent protein (GFP) BM-transplanted mice is demonstrated. Methods and algorithms for nonrigid registration of time-lapse images are introduced, which allows for long-term tracking of cell dynamics over several months.

Keywords In vivo microscopy, Multimodal, Skin regeneration, Bone marrow cells

1 Introduction

Bone marrow (BM)-derived stem cells have captured increasing interest from scientists because of their important functions in tissue regeneration and repair (1, 2). In addition to the adult stem cells, such as mesenchymal (MSC) and hematopoietic (HSC) stem cells, a large variety of immune cells derived from BM have crucial roles and functions in regulating the immune system during a peripheral infection or inflammation (3, 4). Especially in skin biology,

BM-derived cells not only contribute to aid skin regeneration by differentiating into keratinocytes but also are actively involved in the immune response in the form of surveillance and inflammation (5–7). A better understanding of the functions and dynamics of BM-derived cells will necessarily foster advances in regenerative medicine and the clinical application of BM transplant-related therapy for a variety of diseases. Considering the complicated behaviors of BM-derived cells, which involve fast cellular dynamics, including differentiation, migration, and interactions with other cells and the microenvironment of natural tissue, a high-resolution real-time 3-D in vitro and in vivo imaging tool is needed to track these cellular dynamics in order to study their functions, roles, and contributions in health and disease.

Histology has been the gold standard for cell biology research. Due to its destructive nature, however, histology can only study cell structure, or potentially cell function, at a fixed time point. This end-point method is thus not suited for real-time imaging and longitudinal studies. By virtue of its high spatial resolution and noninvasive nature, optical microscopy has been conventionally used for in vitro cell tracking. Because the conventional optical microscope does not offer sufficient depth resolution, it only works well for thin samples, and cell tracking with a conventional optical microscope is primarily performed on the basis of a 2-D culturing environment, for example, in cell culture plates. Confocal microscopy is a 3-D imaging technique that can be used for imaging relatively thick living tissue, and can be operated in both fluorescence and reflectance mode (8). Due to the short wavelength of the excitation beam (normally blue or UV light), confocal microscopy can be problematic due to photodamage and is limited by a relatively short penetration depth. Fortunately, these problems can be overcome with multiphoton microscopy (MPM) which utilizes longer wavelength excitation (normally in the near-infrared) and achieves high 3-D resolution based on nonlinear optical effects, such as two-photon excited fluorescence (TPEF) (9) and second harmonic generation (SHG) (10). High penetration depths of over 1 mm have been achieved with 1.3 µm excitation light in brain tissue (11). Presently, intravital cellular imaging with TPEF is routinely performed. Optical coherence tomography (OCT) is another promising noninvasive imaging technology that is capable of 3-D visualization (12). OCT constructs depth-resolved images by using interferometric techniques to measure the time-of-flight of scattered photons (13). OCT generates images based on the light-scattering properties of samples, making it a noninvasive technique. However, OCT lacks molecular contrast which can be helpful for assessing the functional states of cells and tissues.

As any single imaging technology has specific advantages and limitations, multimodal imaging methods that provide complementary information are desirable. Various multimodal methods have been developed for real-time imaging, such as integrating MPM and OCT (14), coherent anti-Stokes Raman scattering (CARS) and OCT (15), CARS and MPM (16), and reflective confocal and MPM (17, 18), among others. In particular, the combination of optical coherence microscopy (OCM (19, 20), the high-resolution variant of OCT) and MPM has drawn considerable attention as this combination can provide co-registered structural and molecular information of the sample, and can be based on a single laser source (21–24). In this configuration, OCM provides structural information and MPM provides molecular or functional information about the sample. The two modalities are thus complementary, and enable a new method to visualize the dynamic behavior of cells with high 3-D resolution in complicated microenvironments.

This chapter will focus on imaging and tracking BM-derived cells in in vivo mouse skin with a combined MPM/OCT microscope. We show the experiments and results on short- and long-term tracking of GFP-labeled cells derived from transplanted BM, and their responses and functions in the processes of wound healing and skin grafting. The technologies described in the chapter include imaging the collagen structure with SHG, morphological changes and migration of GFP-labeled immune cells with TPEF, and structural information and microvasculature regeneration with OCT/OCM. Experimental procedures and protocols associated with the use of GFP-labeled BM transplants and skin grafts are detailed. An algorithm for image registration with nonrigid samples (such as soft tissues, including skin) is included as well, which enables long-term tracking of cellular dynamics in dynamic tissue microenvironments. The technical and experimental information detailed in this chapter will provide guidance to researchers interested in this topic and in performing related research.

1.1 Multiphoton Microscopy

MPM is a high-resolution imaging technique that is based on nonlinear optical processes, such as two-photon excitation fluorescence (TPEF) imaging (9) and second harmonic generation (SHG) imaging (10). TPEF imaging is based on the absorption of two near-infrared photons and the subsequent emission of a single photon in the visible wavelength range. TPEF can work both for exogenous and endogenous fluorophores. SHG imaging is based on the process of frequency doubling by which two near-infrared photons are converted into a single photon with exactly twice as much energy as the input photons. SHG signals arise from symmetry properties of molecules, which in biological imaging, is most often present in collagen (25).

Due to the nonlinear processes involved in MPM, a higher density of photons is required, and thus, an ultrashort-pulse laser such as a femtosecond titanium-sapphire laser is typically used in combination with a high numerical aperture (NA) objective lens. The high intensity requirement of MPM also reduces photobleaching that is often common in confocal microscopy (CM), and restricts both TPEF and SHG signals from being generated away from the focal volume. This eliminates the need for spatial filtering that is essential for CM. As well, the use of near-infrared excitation wavelengths results in increased penetration depth due to the low absorption and scattering from tissue in this wavelength range. As a result of these properties, MPM presents many advantages over many traditional optical imaging techniques for the acquisition of real-time cellular level images from biological samples (26).

1.2 Optical Coherence Tomography (OCT)

Optical Coherence Tomography (OCT) is a noninvasive scattering-based imaging technique used to obtain high-resolution cross-sectional images of tissue samples (12). OCT operates on the principle of optical ranging by which time-of-flight information from scattered photons is obtained to determine the depth from which they were scattered. This is very similar to ultrasound imaging except that OCT imaging uses optical radiation as opposed to acoustic waves. As the speed of light is much faster than can be accurately detected by direct measurement, low coherence interferometry, requiring the use of a broad bandwidth light source, is employed to acquire the time-of-flight information from the sample. In this scheme, light is split into a sample and reference arm and light scattered from the sample is recombined with the reference light at the detector to form an interference pattern, from which the depth information in the sample is encoded. The beam in the sample arm is scanned in both transverse directions in order to obtain a full 3-D OCT volume. In OCT, the axial resolution is determined by the bandwidth of the source and is often on the order of a few microns, while the transverse resolution is determined by the size of the focused beam and is typically on the order of 10 μm.

Optical coherence microscopy (OCM) is a variant of OCT that utilizes a high NA lens to achieve higher transverse resolution (19, 20). This, however, comes at the expense of a shorter depth-of-field due to the confocal gating of the high NA lens, allowing *en face* planes to be obtained instead of a depth-resolved cross-sectional image as in OCT. The confocal gating provided by the high NA lens combined with the coherence gating provided by the low coherence interferometry used in OCM provides a higher rejection of out-of-focus photons when compared with reflectance confocal microscopy, thus allowing for significantly deeper imaging depths in highly scattering tissues (20).

In addition to measuring scattering signals in tissue, OCT can also be used to visualize the microvasculature in living tissue. This is typically performed by measuring the time-dependent changes in the scattering signal from the sample (27–29). The signal inside blood vessels will be changing over time, and at a much higher rate than outside of the vessels where the scatterers are relatively static. One such method used to visualize the microvasculature in live tissue is known as phase-variance OCT (30, 31). Phase-variance methods are based on making several measurements at a given location and computing the changes in the signal over time. In areas of greater change, a higher signal will be generated. Phase-variance methods are of particular use in assessing the state of skin as it undergoes repair, allowing the microenvironment with which BM-derived cells interact to be better visualized and understood, particularly in skin grafting and wound healing studies.

1.3 Time-Lapse Imaging of Nonrigid Tissue

For long-term time-lapse imaging in skin, registration of consecutive images is critical for tracking bone marrow-derived cells. A challenge is presented when the sample to be imaged undergoes significant mechanical deformations. This can occur in normal daily variations of skin position, but most problematically in situations such as wounding or grafting of the skin, where these deformations cannot be accounted for by traditional linear registration techniques. Instead, nonrigid image registration methods may be employed to better account for these deformations (32). Nonrigid registration algorithms utilize physical landmarks that present themselves with unique spatial relationships that are preserved between imaging sessions. Using these spatial relationships, a transformation grid can be applied to align images at different time points by nonrigid warping of the image. In skin, this method can easily be implemented by using the SHG signal from collagen. By exploiting the spatial relationships between adjacent hair follicles, improved image registration can be readily obtained through nonrigid transformation in wound healing and skin grafting studies.

1.4 GFP Bone Marrow-Derived Cells and Their Function in Regeneration of Skin

BM contains stem and immune cells which are known to play key roles in regeneration and repair of many tissue sites (1, 2). BM stem and progenitor cells consist of hematopoietic stem cells (HSCs) and mesenchymal stem cells (MSCs). HSCs have the potential to differentiate into all blood cell lineages and reconstitute the entire hematopoietic and immune systems following transplantation in vivo. MSCs can differentiate into osteocytes, adipocytes, chondrocytes, hepatocytes, myocytes, cardimyocytes, neurons, and keratinocytes as well. In addition to these predominant stem cells, BM-derived cells have been found to be actively involved in immune processes (3, 4, 33). In skin regeneration, BM-derived cells contribute not only by differentiating into keratinocytes but also by immune surveillance and inflammation, functions which are commonly found in skin wound healing and following skin grafting (5–7).

Since the most efficient and widely used intravital microscopy techniques are based on fluorescence, we used BM from GFP-expressing mice. GFP is a very important in vivo biomarker for cellular imaging that fluoresces at the peak of 509 nm when excited with blue or UV light (34, 35). GFP has a large two-photon absorption coefficient, which makes it well suited for cellular imaging with MPM. GFP can be readily transfected into a variety of cell types, and transgenic mice can express GFP constitutively in all of its cells. Therefore, a wild-type host mouse with transplanted BM from a GFP-transfected mouse is a good model for tracking the dynamics of different types of cells originating from the transplanted BM and expressing GFP, all within a non-GFP-expressing tissue environment of a wild-type mouse.

2   Materials

2.1   Wild Type Mice (Recipients)

6–10-week-old female wild-type C57BL/6 mice.

2.2   GFP Transfected Mice (Donor)

Male transgenic mice with global GFP expression (C57B/6-Tg (CAG-EGFP) 10sb/J).

2.3   Materials and Tools for BM Transplant and Skin Wounding

1. Isoflurane anesthesia gas source.
2. Oxygen source.
3. 70 % ethanol.
4. Sterile phosphate buffered saline (PBS).
5. ACK Lysis buffer.
6. DMEM solution.
7. 100 × 20 mm plastic petri dishes.
8. Mouse nose cone.
9. 10 % povidone/iodine.
10. Sterile gauze pads.
11. 3/4 in. sheer Band-Aid-type sheer bandage (CVS, Curad).
12. Microdissecting forceps (4 in., half-curved, serrated; 5 in., straight, serrated or non-serrated).
13. Microdissecting scissors (4½ in., curved, with very sharp points; 4 in., straight).
14. Watchmaker's or jeweler's forceps (4½ in., 0.17 mm wide, 0.10 mm thick; Dumont no. 7).
15. Scalpel blades (Bard Parker no. 10).
16. 10 ml syringes (Luer Lock).
17. 263/8 G needles.
18. 40 μm filter paper.
19. Stereo microscope.

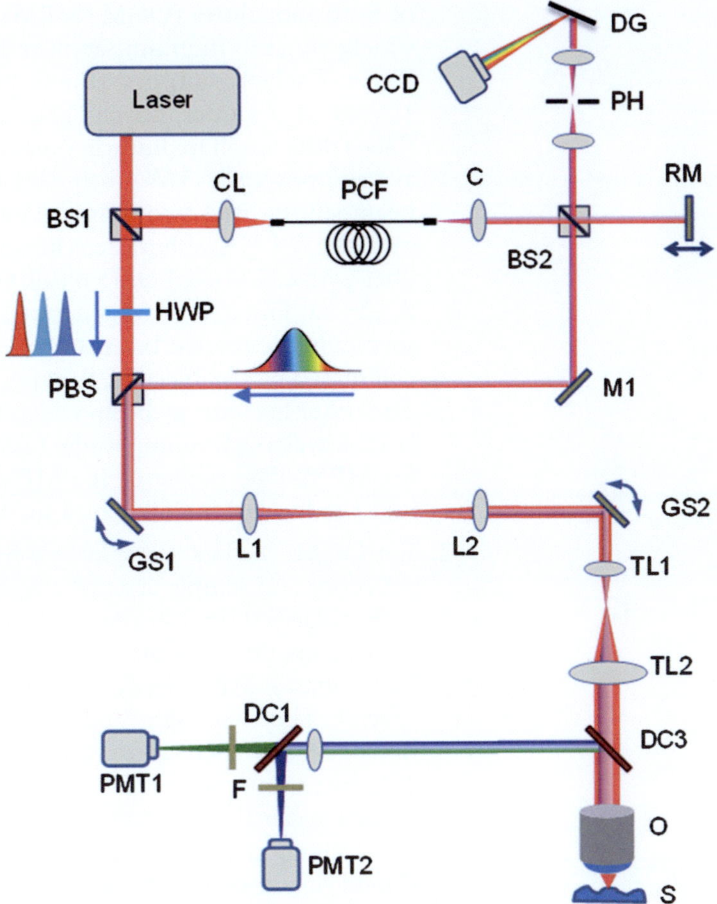

Fig. 1 Schematic of the integrated optical microscope. Abbreviations: *BS* beam splitter; *C* collimator, *CCD* charge-coupled device line-scan camera, *CL* coupling lens, *DC* dichroic mirror, *DG* diffraction grating, *F* filter, *GS* galvanometer scanner, *HWP* half-wave plate, *L* lens, *O* objective, *M* high reflection mirror, *PCF* photonic crystal fiber, *PBS* polarizing beam splitter, *PH* pinhole, *PMT* photo-multiplier tube, *RM* reference mirror, *S* sample, *TL* telescope lens (24)

2.4 Integrated Microscope

The schematic of the integrated OCM/MPM microscope is shown in Fig. 1. This system combines OCT (including the variant functions of OCM and phase variance detection for microvasculature imaging) and MPM (including SHG and TPEF). The multiple imaging modalities are integrated into a single microscope frame, and are thus able to generate spatially co-registered multimodal images.

1. The source of the integrated microscope is based on a single laser but with a specially designed dual-spectrum configuration (36). This design is advantageous in that the single laser enables a more robust and reliable system, while the separately controlled dual-band configuration allows for optimal performance

of both modalities (OCM/MPM). The laser is a high-power, widely tunable titanium:sapphire laser (Mai-Tai HP, Spectra-Physics) which outputs 100 fs pulses with a bandwidth of 10 nm at a center wavelength tunable within the range of 730–1000 nm. The linearly polarized output (with maximum average power of 3 W) from this laser is divided by a 90/10 beam-splitter into two beams. The higher power beam is coupled by a 0.4 NA aspheric lens into a 0.5 m long photonic crystal fiber with a NA of 0.1 and a mode field diameter of 7 μm (LMA-8, Crystal Fiber A/S) to generate super-continuum (SC). This spectrally broadened beam with a bandwidth up to 150 nm is collimated by an objective, and used as the OCT/OCM source. The other low energy beam which has a narrow bandwidth but large wavelength tuning range functions as the excitation light for MPM. The power of the MPM and OCM sources can be independently controlled by a set of neutral density filters.

2. For OCM, the beam is split by a 50/50 beam splitter into the reference and sample arms of a Michelson interferometer. Linearly polarized light in the OCM sample arm is rotated 90° by an achromatic half-wave plate and then recombined with the narrowband MPM excitation beam at the polarizing beam splitter. The two collinearly aligned beams are expanded by a telescope and focused by a microscope objective (20x, 0.95 NA, water immersion, Olympus, Inc.) onto the sample. The objective can also be changed to a low NA objective or lens for OCT imaging with a larger field-of-view. The sample is positioned on a motorized stage which can translate the sample in three directions (x,y,z). A pair of galvanometers (Micromax 671, Cambridge Technology) positioned before the telescope scan the beam across the sample.

3. For OCT/OCM, the reference arm light is reflected by a plane mirror mounted on a translation stage. The spectral interference pattern of the reference and sample arm beams is detected for OCM acquisition by a spectrometer which is based on a diffraction grating and CCD line camera (P2-22-02k40, Dalsa). The frame rate for the line camera depends on the speed and mode of image acquisition (galvanometer vs. stage) and can be up to 35 kHz. OCT/OCM images are generated after several processing steps including compensating for unbalanced dispersion in the sample arm and for nonuniform distribution of the spectrum on the CCD due to nonlinearity of the diffraction grating. The epi-collected MPM fluorescence signal is diverted by a long-pass dichroic mirror and band-pass filtered. This filter is easily interchanged to detect various fluorescence or second-harmonic generation signals. The fluorescence signals of different spectral band are detected separately by two photomultiplier tubes (H7421, Hamamatsu). Control of the integrated system,

as well as image formation and display, is managed through a Labview interface. For large field-of-view OCM/MPM images, mosaic image acquisition is performed by translation of the motorized stage between single images.

---

3   Methods

3.1   Bone Marrow Transplant

Bone marrow harvested from donor male mice with global GFP expression (C57BL/6-Tg (CAG-EGFP) 10sb/J) (37) is transplanted to female wild-type recipient mice (C57B/6). These GFP BM-transplanted mice are informative animal models for investigating the dynamic behavior of BM-derived cells, which express GFP, and can be clearly visualized and differentiated with MPM. Radiate 6–10 week female wild-type recipient mice (C57B/6) with up to 2 doses of 6 GY (administered 4 h apart).

1. Sacrifice donor male mice with global GFP expression (C57BL/6-Tg (CAG-EGFP) 10sb/J) via CO_2 inhalation.

2. Dissect and clean hind limb bones, and place them in phosphate buffered saline (PBS).

3. Crush dissected bones with mortar and pestle.

4. Strain the crushed solution with 40 μm filter paper and lyse red blood cells with an ACK lysing buffer.

5. Count and dilute cell concentration to approximately 7×10^6 cells/ml. Keep the solution on ice prior to transplant.

6. Transplant the prepared bone marrow cells (150 μl, 10^6 cells) into irradiated wild-type mice by tail vein injection.

3.2   In Vivo Multimodal Skin Imaging of GFP BM-Transplanted Mice

In vivo skin imaging with the integrated multimodal microscope is performed in GFP bone marrow transplanted mice.

1. A 6–10-week-old wild-type mouse is anesthetized with isoflurane gas and placed on a heating pad.

2. Shave the area of skin to be imaged with an electric clipper. Remove any residual hair using surgical scissors and a stereo microscope. Try to remove as much hair as possible without injuring the skin (Note 1).

3. Position the anesthetized mouse on the motorized stage of the multimodal microscope and hold it in place by gently clamping the skin. The skin site to be imaged will be pressed against a coverslip.

4. Apply drops of glycerol to the skin as an index-matching agent.

5. MPM and OCT imaging of the skin is performed through the coverslip. Multimodal images of wounded mouse ear skin are shown in Fig. 2.

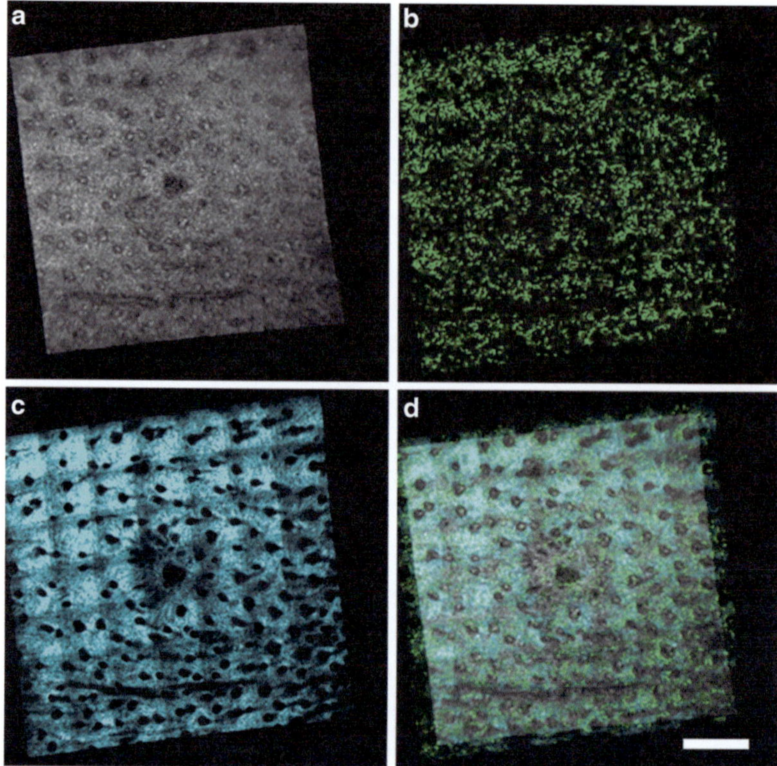

Fig. 2 Multimodal imaging of ear skin in a GFP BM-transplanted mouse 30 days following an excisional wound. (**a**) *En face* structural OCT section showing individual hair follicles. (**b**) Wide-area TPEF mosaic of the GFP expressing BM-derived cells in the skin. (**c**) Wide-area SHG mosaic of collagen. The central dark region represents the wound while the hair follicles appear as smaller dark regions. (**d**) *En face image* of the three modalities overlaid. Scale bar is 500 μm

6. OCT images with a large field-of-view are obtained by using a low NA objective (Fig. 2a). OCT images show the structural morphology of skin tissue with a larger penetration depth, since OCT is based on the light-scattering properties of the sample.

7. TPEF images (Fig. 2b) will show the locations of BM-derived GFP cells.

8. SHG images (Fig. 2c) will show the collagen distribution in the dermis.

9. Images from different modalities can be overlaid (Fig. 2d), which shows the spatial co-registration of all the images. Complementary information is present from the overlaid image. A cross-sectional image extracted from the overlaid volumetric dataset shows the depth-dependent information and the sample properties obtained from the different modalities.

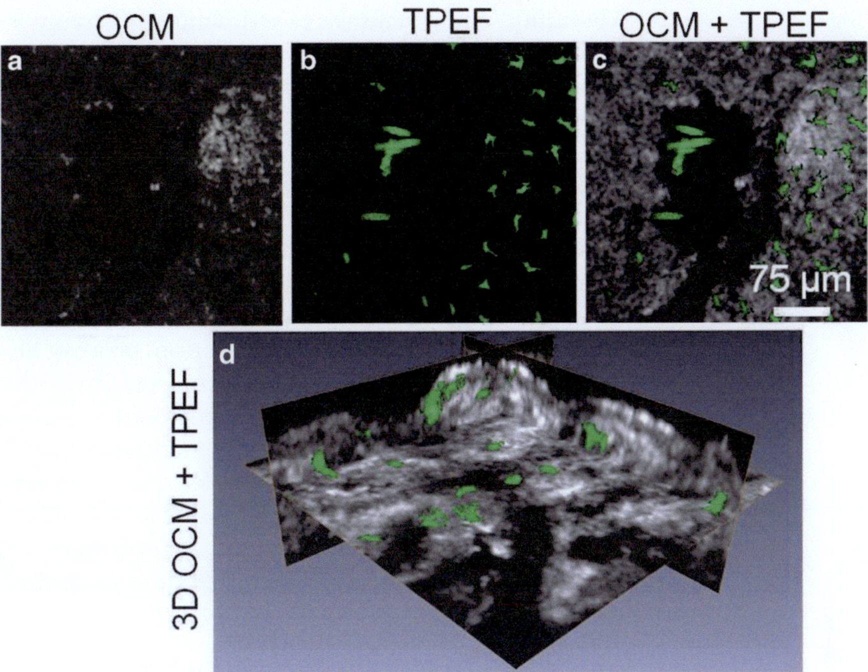

Fig. 3 (**a**) OCM and (**b**) TPEF images of skin from a GFP BM-transplanted mouse. A hair follicle is visible in the OCM image while TPEF visualizes individual GFP cells in addition to autofluorescence from hairs. (**c**) The overlay of the OCM and TPEF images allows the tissue microenvironment of the GFP cells to be seen. (**d**) 3-D orthoslices of the volumetric OCM-TPEF overlays

10. High-resolution OCM images can be obtained using a high NA objective (Fig. 3a).

11. Zoomed-in views of the TPEF images (Fig. 3b) show the detailed morphology of BM-derived GFP cells.

12. Overlaid OCM and TPEF images (Fig. 3c) show the positions of the GFP cells in 2-D.

13. 3-D orthoslices of the volumetric OCM-TPEF (Fig. 3d) overlays reveal the locations of GFP cells in the 3-D background of the skin structure.

3.3 Skin Wounding

A skin wound is made on a GFP BM-transplanted mouse for real-time imaging and tracking of the dynamics and responses of BM-derived cells in wound healing processes.

1. Anesthetize the BM-transplanted mouse with isoflurane gas.

2. Shave the skin (either dorsal or ear) with electric clippers.

3. Carefully remove remaining hair from the region to be imaged using surgical scissors and a stereo microscope.

4. Wound the skin site to be imaged by taking either a 1 mm diameter punch biopsy or making a 1 mm long linear incision manually with a fine-tipped scalpel.

5. Wounding and imaging experiments preferably should be performed at least 2 months after the BM transplant, when complete marrow engraftment and bone marrow cell production has resumed.

3.4 Nonrigid Co-registration Algorithm

Natural skin is a highly flexible organ which experiences significant mechanical distortion during daily activities as well as contraction during wound healing. In this case, a nonrigid image registration process must be introduced to account for the nonrigid changes during wound healing in order to track the same skin site over several months. This nonrigid registration process also enables one to separate the small-scale movements and dynamics of single cells from the large-scale changes due to tissue deformation. The nonrigid registration algorithm in this method is based on using the locations of hair follicle as landmarks. The registration procedure consists of detecting hair follicle positions, initializing a follicle matching algorithm, performing a grid transformation, and iteratively optimizing the solution. The different steps of the algorithm are illustrated in Fig. 4.

1. Choose two *en face* images from two volumetric skin image datasets that were acquired from two time points, and are to be registered (Fig. 5a and b). The images can be either SHG or OCM structural images that have good contrast between hair follicles and dermis (Note 2).

2. Filter the images with a difference-of-Gaussian (DOG) kernel at different scales. Local maxima in the DOG image are detected following the application of a threshold. These maximum positions are considered as the center locations of the follicles (Note 3).

3. Evaluate the match quality of each hair follicle in one image to each follicle in the other image to be co-registered. This is done by selecting a fixed number of neighboring follicles (typically 10–15) and calculating the matching quality between the two

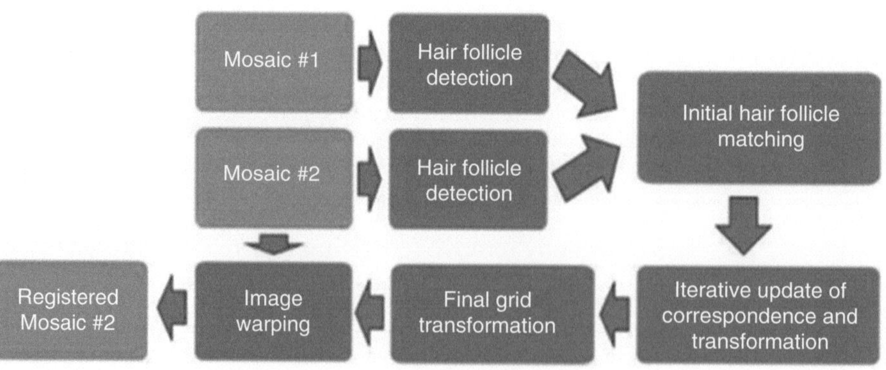

Fig. 4 Block diagram of the nonrigid image registration algorithm

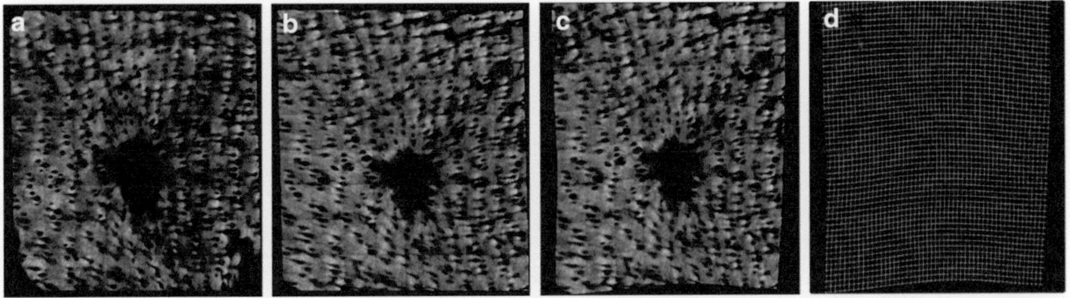

Fig. 5 SHG images of mouse skin on (**a**) day 1 and (**b**) day 2 after wounding. (**c**) Warped image after the nonrigid image registration process. (**d**) Warping grid of image shown in (**c**)

sets of neighboring points after local rotation and scaling operations. This method assumes that the local deformation within the range of the neighboring points can be approximated by a rigid transformation.

4. Determine matched follicles that have a higher match quality than a given threshold. Use the positions of those matched follicles to define an initial transformation between the two images.

5. Match remaining follicles by an iterative procedure that adds new follicle matches to the transformation model each time a match is determined.

6. Interpolate the transformation function to determine a nonrigid warping transformation matrix and 2-D warping grid (Fig. 5d).

7. Warp the original image based on the warping grid to obtain a registered image (Fig. 5c).

3.5 Time-Lapse Imaging and Tracking of BM Cells in Would Healing

The combination of multiple imaging modalities allows in vivo visualization of BM-derived cell dynamics and their function in wound healing processes. In addition to the short-term cell migration dynamics, the nonrigid image registration enables fundamental dynamics of BM-derived cells to be tracked over a period of several months.

1. A full-thickness excisional wound is made in the dorsal skin of a GFP BM-transplanted mouse, following the protocols of Section 3.3.

2. Following the wound, MPM images are acquired at several time points over 45 minutes on the first day of wounding (Fig. 6) to reveal the short-term dynamic behavior of BM-derived cells.

3. SHG images provide the structural information of the skin environment around the wound site as well as provide guidance for image registration to track cell dynamics.

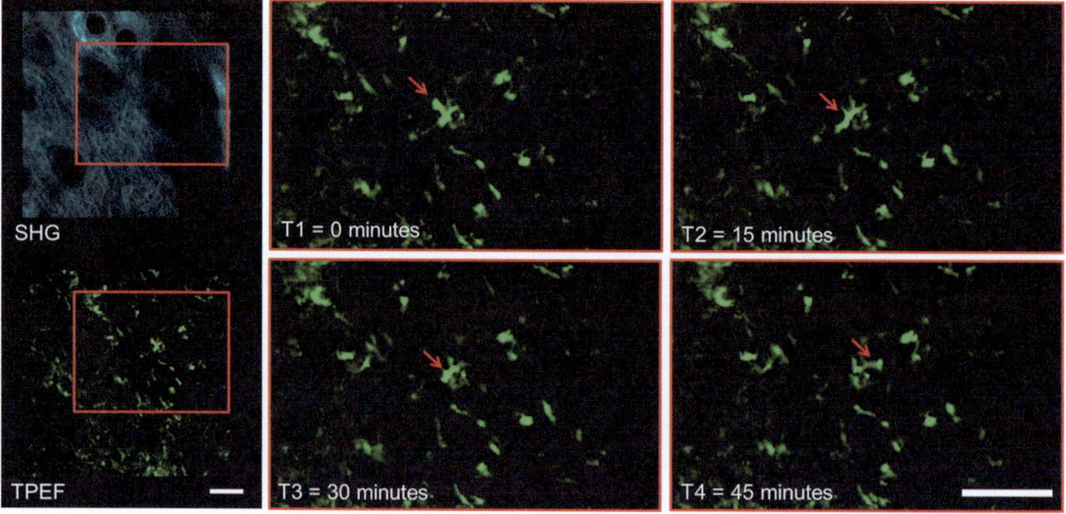

Fig. 6 Time-lapse imaging of dendritic cells in vivo, showing their activation over 45 minutes following a cutaneous wound. Magnified view of the cell cluster indicated by the *red box* in the wide-area SHG and TPEF images, showing the morphological change and migration of a dendritic cell in response to the wound. This behavior is consistent with the known function of Langerhans cells and suggests that cells may be in the process of migrating from the skin to the lymph nodes. Scale bars are approximately 100 μm

4. Time-lapse images reveal dynamics of a cluster of Langerhans cells in the vicinity of wound site within 24 h (Fig. 6). These cells are identified on the basis of their unique morphology and location within the epidermis of the skin (Note 4) (38).

5. High-magnification TPEF images (Fig. 6) show the detailed morphology of the GFP BM-derived cells. The cells transform their morphology from one having many extensions for sensing the extracellular environment for antigens, to one of an amorphous shape suitable for migration. A magnified view of the cell cluster shows the morphological change and the eventual disappearance of many Langerhans cells in response to the wound. This behavior is consistent with the known function of Langerhans cells and suggests that many of the cells have migrated from the skin wound site to the locoregional lymph nodes.

6. Following the short-term (24 h) imaging sequence, OCM/MPM images of the wound site are acquired every 24 h for 2 weeks (Fig. 7).

7. In the GFP/SHG overlays (Fig. 7a–e), the GFP fluorescence channel shows the dynamic changes in the BM-derived cells. The most notable effect is the large increase in the number of GFP cells around the wound site on day 2 (Fig. 7b). Based on the proximity of these large cell clusters around the edge of the wound and the timing of their arrival and disappearance, it is likely that these cells are involved in the natural inflammatory and immune responses.

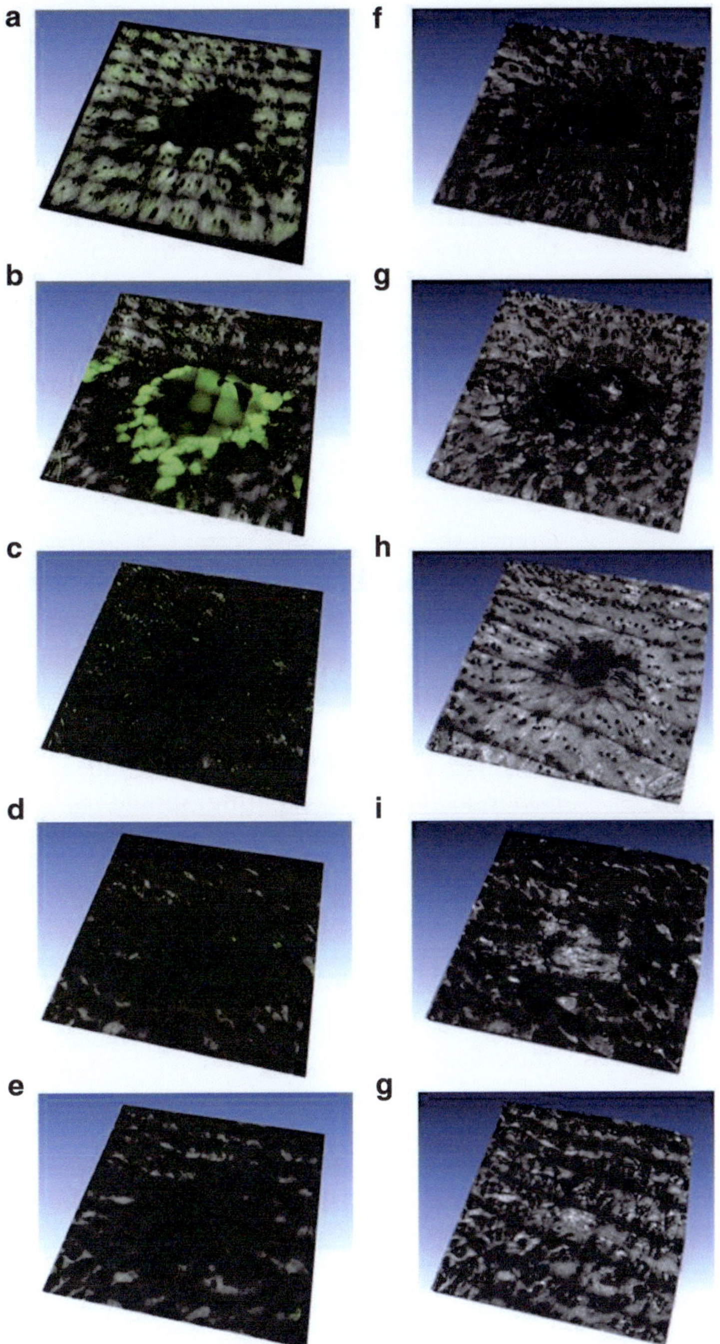

Fig. 7 Three-dimensional (**a–e**) MPM fluorescence images overlaid on SHG and (**f–j**) OCM renderings of the wound site at multiple time points during healing. In (**a–e**), the GFP is represented by the *green channel* and the SHG data is represented in *grey scale*

8. Based on the SHG and OCM signals, the wound site is initially visible as a round hole and the wound gradually heals over 2 weeks (Fig. 7j), as shown by the OCM images. However, the SHG signal in the center of the wound fails to reappear by the end of the 2 week period (Fig. 7e), and the scattering signal (shown by the OCM images) from the wound remains different than the surrounding areas. These results suggest that while the wounded skin has healed, the wound is still lacking the organized collagen found in normal skin.

3.6 Skin Graft

Skin grafts from wild-type donors to GFP BM-transplanted recipient mice are performed following a similar protocol described in ref. (39).

1. Sacrifice a wild-type C57BL/6 donor mouse by CO_2 inhalation.

2. Drench the entire mouse with 70 % ethanol.

3. Dissect and clean ears, and place them in PBS.

4. Surgically remove a 7×7 mm^2 area of donor ventral ear skin and place it in cold PBS. The harvested skin will consist of both the epidermis and dermis of the skin, but not the underlying cartilage.

5. Anesthetize the GFP BM-transplanted recipient mouse with isoflurane gas. Skin grafting should be performed at least 8 weeks after BM transplant.

6. Shave the host skin site for the graft bed (either back or ear) with electric clippers.

7. Surgically remove a 7×7 mm^2 piece of ventral skin from the prepared site on the host using forceps and dissecting scissors (Note 5).

8. Immediately place the skin graft from the donor onto the prepared graft bed of the host.

9. Cover the graft site with petrolatum gauze and wrap the mouse with a strip of elastic bandage.

10. Monitor the integrity of the bandage at least daily, and remove the bandage following 7 days of recovery.

3.7 Time-Lapse Imaging of BM Cells After Skin Grafting

Multimodal images of the skin graft are acquired at different time points after the removal of the bandage. Because the skin graft is from a wild-type mouse, any GFP cells present within the graft area are from host BM-derived cells and their dynamic behavior will be visualized by the TPEF images (40).

1. SHG images are acquired from the graft site, which provide the structural information of the skin environment of the graft as well as provide guidance for image registration to track the cell dynamics (Fig. 8a).

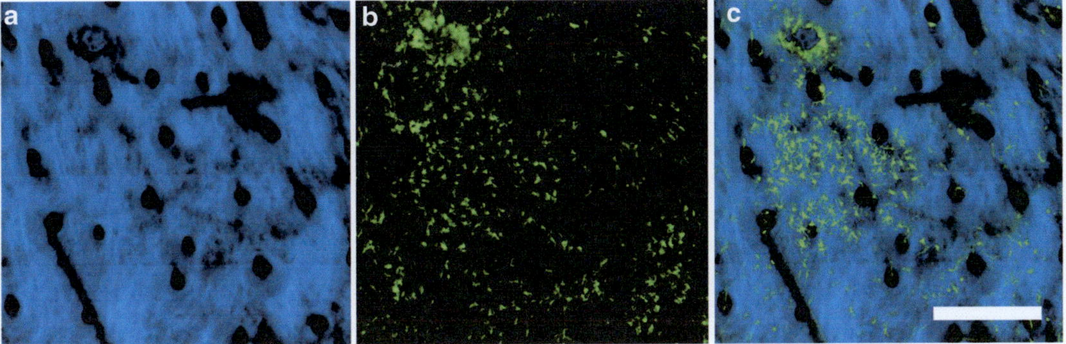

Fig. 8 Axial projections of in vivo volumetric SHG and TPEF images of grafted skin in GFP bone marrow transplanted mice. (**a**) The SHG projection visualizes the collagen network while (**b**) the TPEF projection visualizes the bone marrow-derived cells present throughout the depth of the skin. (**c**) An overlaid projection with high SHG opacity allows the GFP cells in the epidermis to be visualized while masking the GFP cells in the dermis. The presence of a focal cluster of epidermal GFP cells is apparent in the skin graft (**c**). Scale bar is 300 μm. Figure modified from ref. (40) Copyright Wiley-VCH Verlag GmbH & Co. KGaA. Reproduced with permission

2. TPEF images are acquired, which show the morphology and dynamics of BM-derived GFP cells (Fig. 8b).

3. SHG and TPEF images are overlaid to show the position and dynamics of the BM-derived GFP cells. The presence of clusters of GFP cells in the epidermis of the grafted skin is observed (Fig. 8c). The isolated nature of this cluster suggests that a local precursor cell was derived from the BM and then proliferated to form this cluster of cells. Based on the morphology of these cells, it is likely that they are Langerhans cells.

4   Notes

1. The two-photon-generated autofluorescence background from hair shafts is subtracted from volumetric TPEF data using a segmentation procedure. This was achieved by applying a threshold to the volume, identifying isolated features, and assessing them based on their size, aspect ratio, and length. The inverse of the resulting binary mask was multiplied with the original image volume to remove the hair shafts. Although this method does not account for weaker autofluorescent features, it enables the removal of the largest features which can affect the interpretation of the images.

2. While the registration is performed in two dimensions, it is possible to correct the axial dimension as well by computationally shifting the SHG volumes along the axial dimensions to make the surface of the dermis flat.

3. Local minima in the scale-space are selected as possible feature points. The initial set of possible hair follicle points that is found is very high and contains many incorrect matches. Points are eliminated based on thresholding the DOG images and an additional metric that evaluates the radial symmetry of the local image gradients, resulting in a more accurate estimate of the follicle positions. This process is performed on SHG images from multiple depths, and matches of the same follicle at different depths are merged to give the final estimate of the hair follicle locations.

4. Langerhans cells (LC) are a class of dendritic cells whose primary function is to sense the extracellular environment for the presence of antigens. These cells reside in the epidermis of the skin and are typically immobile under steady-state conditions. Upon detection and uptake of antigens, these cells migrate to the locoregional lymph nodes to present antigens to other cells in the immune system as part of the adaptive immune response. Under steady-state conditions, LCs have a very distinct morphology which consists of dendrites that probe the local microenvironment for antigens. Upon activation, these dendritic processes are retracted, resulting in an amorphous morphology that enables migration of the LCs to the lymph nodes. Understanding the functional role of LCs is an active area of research. While these cells were initially thought to play a role in initiating immune responses in reaction to foreign antigens, recent studies have suggested their primary function may be to support immune tolerance (41). The time-lapse imaging shows that a large number of these cells undergo a reorganization of their morphology over a 45 minutes time period and migrate from their initial locations. This is, to our knowledge, the first in vivo time-lapse imaging of the activation of individual Langerhans cells.

5. Using watchmaker's forceps, pinch the skin where the topmost horizontal cut will be made, which is at the top of the scapula, just lateral to the spine. Rest the blades of the dissecting scissors flat on top of the pinched skin and cut skin as superficially as possible. Take care not to angle the scissors downward when cutting or the panniculus carnosus will be cut. If the panniculus is accidentally removed, the skin will not engraft. If the cut is not sufficiently deep to reach the panniculus, repinch the skin and carefully cut to the level of the panniculus. There should be no white or pigmented dermal tissue present in the cuts, as this will interfere with removal of the intact skin. Complete a square with four cuts. Using watchmaker's forceps, grasp a corner of the square skin graft and gently pull it off the skin bed.

Acknowledgments

We thank Dr. Marina Marjanovic and Dr. Michael De Lisio for their helpful discussions and assistance with this research, and Mr. Darold Spillman for his technical, logistical, and information-technology support to this research. We also thank Eric Chaney and Ziad Mahmassani for their laboratory assistance with our biological resources. This research was supported in part by grants from the National Science Foundation (CBET 08–52658 ARRA, CBET 10–33906, S.A.B). Benedikt Graf was supported by the Pre-doctoral National Institutes of Health Environmental Health Sciences Training Program in Endocrine, Developmental and Reproductive Toxicology at the University of Illinois at Urbana-Champaign. Additional information can be found at http://biophotonics.illinois.edu.

References

1. Bajada S, Mazakova I, Richardson JB, Ashammakhi N (2008) Updates on stem cells and their applications in regenerative medicine. J Tissue Eng Regen Med 2:169–183

2. Weissman IL (2000) Stem cells: units of development, units of regeneration, and units in evolution. Cell 100:157–168

3. Fujisaki J, Wu J, Carlson AL, Silberstein L, Putheti P, Larocca R, Gao W, Saito TI, Lo Celso C, Tsuyuzaki H, Sato T, Cote D, Sykes M, Strom TB, Scadden DT, Lin CP (2011) In vivo imaging of Treg cells providing immune privilege to the haematopoietic stem-cell niche. Nature 474:216–219

4. Nitschke C, Garin A, Kosco-Vilbois M, Gunzer M (2008) 3D and 4D imaging of immune cells in vitro and in vivo. Histochem Cell Biol 130:1053–1062

5. Wu Y, Wang J, Scott PG, Tredget EE (2007) Bone marrow-derived stem cells in wound healing: a review. Wound Repair Regen 15 (Suppl 1):S18–S26

6. Badiavas EV, Abedi M, Butmarc J, Falanga V, Quesenberry P (2003) Participation of bone marrow derived cells in cutaneous wound healing. J Cell Physiol 196:245–250

7. Brittan M, Braun KM, Reynolds LE, Conti FJ, Reynolds AR, Poulsom R, Alison MR, Wright NA, Hodivala-Dilke KM (2005) Bone marrow cells engraft within the epidermis and proliferate in vivo with no evidence of cell fusion. J Pathol 205:1–13

8. Pawley JB (ed) (2006) Handbook of biological confocal microscopy. Springer, New York, NY

9. Denk W, Strickler JH, Webb WW (1990) Two-photon laser scanning fluorescence microscopy. Science 248:73–76

10. Campagnola PJ, Loew LM (2003) Second-harmonic imaging microscopy for visualizing biomolecular arrays in cells, tissues and organisms. Nat Biotechnol 21:1356–1360

11. Kobat D, Durst ME, Nishimura N, Wong AW, Schaffer CB, Xu C (2009) Deep tissue multiphoton microscopy using longer wavelength excitation. Opt Express 17:13354–13364

12. Huang D, Swanson EA, Lin CP, Schuman JS, Stinson WG, Chang W, Hee MR, Flotte T, Gregory K, Puliafito CA et al (1991) Optical coherence tomography. Science 254:1178–1181

13. Schmitt JM (1999) Optical coherence tomography (OCT): a review. IEEE J Sel Top Quantum Electron 5:1205–1215

14. Jeong B, Lee B, Jang MS, Nam H, Yoon SJ, Wang T, Doh J, Yang BG, Jang MH, Kim KH (2011) Combined two-photon microscopy and optical coherence tomography using individually optimized sources. Opt Express 19:13089–13096

15. Hoffmann C, Hofer B, Unterhuber A, Poavzay B, Morgner U, Drexler W (2011) Combined OCT and CARS using a single ultrashort pulse Ti:Sapphire laser. Proc SPIE 7892:78920B

16. Wong CS, Robinson I, Ochsenkuhn MA, Arlt J, Hossack WJ, Crain J (2011) Changes to lipid droplet configuration in mCMV-infected fibroblasts: live cell imaging with simultaneous CARS and two-photon fluorescence microscopy. Biomed Opt Express 2:2504–2516

17. Masters B, So P (2001) Confocal microscopy and multi-photon excitation microscopy of human skin in vivo. Opt Express 8:2–10

18. Choi SH, Kim WH, Lee YJ, Lee H, Lee WJ, Yang JD, Shim JW, Kim JW (2011) Visualization of epidermis and dermal cells in ex vivo human skin using confocal and two-photon microscopy. J Opt Soc Korea 15:61–67

19. Aguirre AD, Hsiung P, Ko TH, Hartl I, Fujimoto JG (2003) High-resolution optical coherence microscopy for high-speed, in vivo cellular imaging. Opt Lett 28:2064–2066

20. Izatt JA, Hee MR, Owen GM, Swanson EA, Fujimoto JG (1994) Optical coherence microscopy in scattering media. Opt Lett 19:590–592

21. Vinegoni C, Ralston T, Tan W, Luo W, Marks DL, Boppart SA (2006) Integrated structural and functional optical imaging combining spectral-domain optical coherence and multiphoton microscopy. Appl Phys Lett 88:053901

22. Graf BW, Boppart SA (2012) Multimodal in vivo skin imaging with integrated optical coherence and multiphoton microscopy. IEEE J Sel Top Quantum Electron 18:1280–1286

23. Wu QF, Applegate BE, Yeh AT (2011) Cornea microstructure and mechanical responses measured with nonlinear optical and optical coherence microscopy using sub-10-fs pulses. Biomed Opt Express 2:1135–1146

24. Zhao YB, Graf BW, Chaney EJ, Mahmassani Z, Antoniadou E, DeVolder R, Kong H, Boppart MD, Boppart SA (2012) Integrated multimodal optical microscopy for structural and functional imaging of engineered and natural skin. J Biophotonics 5:437–448

25. Stoller P, Reiser KM, Celliers PM, Rubenchik AM (2002) Polarization-modulated second harmonic generation in collagen. Biophys J 82:3330–3342

26. Helmchen F, Denk W (2005) Deep tissue two-photon microscopy. Nat Methods 2:932–940

27. Chen ZP, Milner TE, Srinivas S, Wang XJ, Malekafzali A, vanGemert MJC, Nelson JS (1997) Noninvasive imaging of in vivo blood flow velocity using optical Doppler tomography. Opt Lett 22:1119–1121

28. Makita S, Hong Y, Yamanari M, Yatagai T, Yasuno Y (2006) Optical coherence angiography. Opt Express 14:7821–7840

29. Barton JK, Stromski S (2005) Flow measurement without phase information in optical coherence tomography images. Opt Express 13:5234–5239

30. Zhao YH, Chen ZP, Saxer C, Xiang SH, de Boer JF, Nelson JS (2000) Phase-resolved optical coherence tomography and optical Doppler tomography for imaging blood flow in human skin with fast scanning speed and high velocity sensitivity. Opt Lett 25:114–116

31. Vakoc BJ, Lanning RM, Tyrrell JA, Padera TP, Bartlett LA, Stylianopoulos T, Munn LL, Tearney GJ, Fukumura D, Jain RK, Bouma BE (2009) Three-dimensional microscopy of the tumor microenvironment in vivo using optical frequency domain imaging. Nat Med 15:1219–1223

32. Crum WR, Hartkens T, Hill DLG (2004) Non-rigid image registration: theory and practice. Br J Radiol 77:S140–S153

33. Shi C, Jia T, Mendez-Ferrer S, Hohl TM, Serbina NV, Lipuma L, Leiner I, Li MO, Frenette PS, Pamer EG (2011) Bone marrow mesenchymal stem and progenitor cells induce monocyte emigration in response to circulating toll-like receptor ligands. Immunity 34:590–601

34. Naumov GN, Wilson SM, MacDonald IC, Schmidt EE, Morris VL, Groom AC, Hoffman RM, Chambers AF (1999) Cellular expression of green fluorescent protein, coupled with high-resolution in vivo videomicroscopy, to monitor steps in tumor metastasis. J Cell Sci 112:1835–1842

35. Hoffman RM (2005) The multiple uses of fluorescent proteins to visualize cancer in vivo. Nat Rev Cancer 5:796–806

36. Graf BW, Jiang Z, Tu H, Boppart SA (2009) Dual-spectrum laser source based on fiber continuum generation for integrated optical coherence and multiphoton microscopy. J Biomed Opt 14:034019

37. Okabe M, Ikawa M, Kominami K, Nakanishi T, Nishimune Y (1997) 'Green mice' as a source of ubiquitous green cells. FEBS Lett 407:313–319

38. Kaplan DH (2010) In vivo function of Langerhans cells and dermal dendritic cells. Trends Immunol 31:446–451

39. McFarland H, Rosenberg A (2001) Skin allograft rejection. In: Current protocols in immunology. Wiley, New York

40. Graf BW, Bower AJ, Chaney EJ, Marjanovic M, Adie SG, De Lisio M, Valero MC, Boppart MD, Boppart SA (2013) *In vivo* multimodal microscopy for detecting bone-marrow-derived cell contribution to skin regeneration. Journal of Biophotonics

41. Matheoud D, Perie L, Hoeffel G, Vimeux L, Parent I, Maranon C, Bourdoncle P, Renia L, Prevost-Blondel A, Lucas B, Feuillet V, Hosmalin A (2010) Cross-presentation by dendritic cells from live cells induces protective immune responses in vivo. Blood 115:4412–4420

Methods Molecular Biology (2013) 1052: 77–88
DOI 10.1007/7651_2013_18
© Springer Science+Business Media New York 2013
Published online: 17 April 2013

Covisualization of Methylcytosine, Global DNA, and Protein Biomarkers for In Situ 3D DNA Methylation Phenotyping of Stem Cells

Jian Tajbakhsh

Abstract

DNA methylation and histone modifications are key regulatory mechanisms in cellular differentiation, and are skewed in complex diseases. Therefore, analyzing the higher nuclear organization of methylated DNA in conjunction with relevant cellular components, such as protein biomarkers, may well add cell-by-cell-specific spatial and temporal information to quantitative molecular data for the discovery of stem cell differentiation-related signaling networks and their exploitation in the therapeutic reprogramming of cells. The *in situ* fluorescent covisualization of methylated DNA (methylated CG dinucleotides = MeC), global DNA (gDNA), and proteins has been challenging, as the immunofluorescence detection of MeC sites requires thorough denaturing of double-stranded DNA for antigen (methylated carbon-5 of cytosine) retrieval. The protocol we present overcomes this obstacle through optimization of cell membrane permeabilization, acid treatment, and intermediate fixation steps to preserve immunostaining of biomarkers and delineate MeC and gDNA, while conserving the captured three-dimensional (3D) structure of the cells; making it suitable for high-resolution confocal microscopy, 3D visualization, and topological analyses of fixed cultured cells as well as fresh and frozen tissue sections.

Keywords: DNA methylation phenotyping, Multicolor immunofluorescence, Confocal microscopy, Stem cell differentiation, Cell-by-cell analysis

1   Introduction

Since every cell has its unique epigenome, in particular its own DNA methylome, methods that can mirror the DNA methylation status of single cells are in demand for epigenetic analysis in stem cell differentiation and disease pathology (1, 2). In situ methods generally have the advantage over molecular methods in that they are able to simultaneously analyze hundreds of thousands of individual cells nondestructively within their native microenvironment in culture and original tissue: this process is usually performed without the necessity of labor-intensive and costly target extraction and enrichment or amplification steps. In this context, visualizing the spatial nuclear distribution pattern of methylated CG dinucleotides (DNA methylation phenotyping) may help to address the currently poor understanding of the relationship between DNA methylation and its large-scale functional organization in cell nuclei, with implications in stem cell differentiation and disease. The visualization of MeC in individual nuclei could be first realized through the development of a polyclonal (3) and later a monoclonal antibody (4) that specifically recognized the methyl group on carbon-5 of the MeC pyrimidine ring. This step remarkably illustrated the uneven distribution of MeC along metaphase chromosomes of various mammalian sources (5–7) and in cytologically prepared interphase nuclei of human lymphocytes (8). Immunolabeling of MeC-residues prerequisites accessibility of the antibody to its epitope on the single-stranded DNA. This status is achieved by denaturation of double-stranded DNA (dsDNA), for which different techniques have been applied: initially UV-irradiation (5–9) or enzymatic DNA digestion with different endonucleases (10). Modification of the denaturation step using hydrochloric acid (HCl) treatment unmasked further MeC sites in chromosomes (11, 12) leading to the detection of abnormal methylation in the heterochromatin of cells from patients with ICF (immunodeficiency, centromeric region instability, and facial anomalies) syndrome (7) and many types of human cancer (13–18). This step was originally developed for stoichiometric immunological staining of 5-bromo-2-deoxyuridine (BrdU) incorporated into newly synthesized dsDNA (19), a method that has been widely used to study the cell cycle. HCl treatment in combination with formamide and heat denaturation was also reported for the combined use of MeC staining and fluorescence *in situ* hybridization (FISH) in cellular specimens and whole animal embryos (20–23). The use of HCl treatment requires adjustment as there is a trade-off between stoichiometric antigen labeling (BrdU or MeC) and DNA decomposition and loss of structural integrity visualized through interference with binding of DNA intercalating dyes such as DAPI and propidium iodide, as previously reported (24, 25). This issue is also relevant to proteinaceous components of the cell, which can be affected by excessive acid treatment. However, the visualization and quantitation of nuclear MeC patterns relative to the spatial distribution of other classes of nucleic acids such as gDNA and nuclear proteins may elucidate new pathways related to epigenetic mechanisms in cellular differentiation and their malfunction in the etiology and pathology of complex traits. Furthermore, stem cells are gaining momentum as model systems in drug discovery and development (26).

Thus, we present a protocol for the advanced high-resolution and 3D fluorescence covisualization of MeC, gDNA, and proteins in structurally well-preserved mammalian cells. This protocol is based on experiences gathered from numerous fluorescence cell assays, and in particular the analysis of MeC–related chromatin texture with a variety of human and mouse cell types and organ tissues. Verified procedures for preservation of the structural integrity of cells have been previously described (27–30). This protocol enables the quantitative measurement of the aforementioned coexisting phenomena as relevant changes in the spatial intensity distribution of the two types of DNA signals in the nucleus: (a) MeC signals created through immunofluorescence targeting of methylated cytosine and (b) gDNA signals generated by subsequent counterstaining of the same cells with 4′,6-diamidino-2-phenylindole (DAPI). As DAPI intercalates into AT-rich DNA, the main component of highly repetitive and compact heterochromatic sequences, these regions become visible as bright foci with higher DAPI-signal density. Figure 1 illustrates the protocol's workflow. We applied the protocol with two consecutive sets of primary and secondary antibodies. This protocol can be adapted for multiple sets of antibodies applied in combination, or in sequential staining steps prior to acid treatment and MeC staining. Key considerations, when designing an experiment to visualize MeC, gDNA, and protein markers in 3D preserved mammalian cells, are explained below.

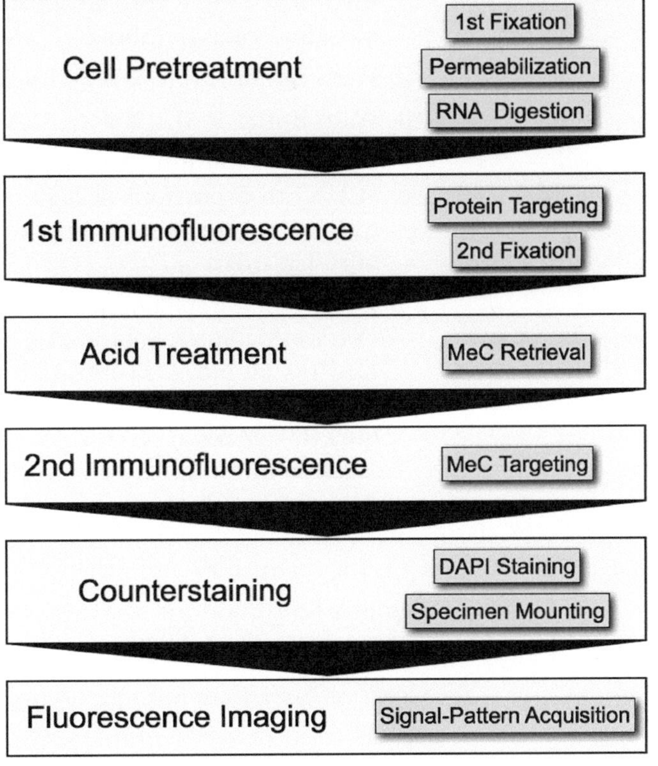

Fig. 1 Workflow of in situ DNA methylation phenotyping assay and imaging: comprising the six major steps (*yellow boxes*) and substeps (*blue boxes*)

2   Materials

Prepare all solutions using deionized water (18 M Ω cm at 25 °C) and analytical grade reagents. With the exception of phosphate buffered saline (room temperature), store all reagents at 2–8 °C (unless indicated otherwise).

2.1   Solutions

1. Phosphate buffered saline (PBS): Prepare 10× concentrated stock solution, a mixture of 1.37 M NaCl, 27 mM KCl, 43 mM Na_2HPO_4, 14.7 mM KH_2PO_4 in water. Dissolve 80 g of NaCl, 2 g of KCl, 14.4 g of Na_2HPO_4, and 2.4 g of KH_2PO_4 in 800 ml of water. Then adjust pH to 7.4, and adjust volume to 1 l. 1× PBS solution can be obtained by dilution of the stock solution.

2. Fixation Solution: 4 % paraformaldehyde w/v in PBS.

3. Permeabilization Solution: 0.5 % w/v saponin, 0.5 % v/v triton X-100 in PBS.

4. RNase Digestion Solution: 100 µg/ml RNase A in PBS.

5. Blocking Solution: 3 % w/v bovine serum albumin fraction V (BSA) in PBS.

6. All primary and secondary antibody solutions are prepared with blocking solution (above).

7. Wash Buffer A: 0.1 % w/v BSA, 0.1 % v/v tween 20 in PBS.

8. Wash Buffer B: 0.1 % w/v BSA in PBS.

9. 2 N HCl Solution: Add 1 portion of 37 % hydrochloric acid (12 N) to 5 portions of water (caution: In this sequence and not reverse, to prevent any strong exothermic reaction with burn hazards!).

10. DNA Counterstaining Solution: Prepare a 5 mg/ml (14.3 mM) Stock Solution by resuspending 10 mg lyophilized DAPI (Cat. No. D21490, Life Technologies) in 2 ml of water. Store stock solution at −20 °C (can freeze and thaw frequently). Make a 143 nM Working Solution by adding 5 µl of stock solution to 50 ml of PBS, and keep (for no longer than 2 weeks) at 2–8 °C in the dark.

2.2   Equipment

CO_2 incubator for mammalian cell culturing

Laminar flow hood for sterile processing of mammalian cells

Oven (for incubations at 37 °C)

Confocal laser scanning microscope

Flat-bottom 12-well culture microplate (3.5 cm^2 well bottom area)

Glass coverslips (18 mm circle, thickness No. 1)

Glass slide (25 × 75 mm)

Whatman™ 3MM CHR Chromatography Paper

3 Methods

3.1 Pretreatment of Cells

This protocol has been optimized for applications with 12-well microplates ($\varnothing = 3.5$ cm). We recommend adjusting volumes and reagent concentrations when using other size dishes. The protocol is suitable for cells cultured on coverslips and either fresh or originally frozen tissue sections (30–50 µm thick). Tissue sections can be processed as floating specimens in microwells (see below). As the immunofluorescence assay part is performed with standard Society for Biomolecular Screening (SBS) format microplates, it is amenable to automation.

1. Culture cells on round cover slips (18 mm, No. 1, 1 oz) in 12-well microtiter plate, using ambient conditions: Usually 37 °C and 5 % CO_2 (see Note 1).

2. Wash cells gently three times with 2 ml of PBS at room temperature. Be careful and avoid pipetting PBS directly onto cells, which may cause detachment of unfixed cells from the coverslip bottom. Add the solution carefully to the sidewall of the culture wells. Also for liquid removal, tilt the dish slightly, so that the solution can be gently taken up from a corner of the well.

3. Fix cells in 1 ml of Fixation Solution for 10–15 min at room temperature, then wash cells three times with 2 ml of PBS (5 min each) (see Note 1). Then add 1–2 ml 0.02 % NaN_3/PBS to cells. At this stage fixed cells can be stored at 2–8 °C for up to several weeks. When ready for the next step, briefly rinse coverslip/tissue in 2 ml of PBS.

4. Treat cells with Permeabilization Solution (1 ml) for 20 min at room temperature, and wash three times with PBS (2 ml for 5 min each) (see Note 2).

5. Treat cells with RNase Digestion Solution (0.6 ml) for 30 min at 37 °C, and wash three times with PBS (2 ml for 5 min each). This will highly degrade tRNAs, which may also carry methylated cytosine molecules, and prevent nonspecific detection (see Note 3). The minimal volume for costly reagents such as enzyme and antibody solutions should be 0.6 ml to ensure the cells being sufficiently covered at any time during incubations of up to 2 h at 37 °C and overnight at 4 °C, as evaporation occurs. This minimum volume provides satisfactory reaction kinetics as experienced and confirmed by subsequent fluorescence imaging.

6. Block nonspecific sites in cells with Blocking Solution (1 ml) for 30 min to 1 h at 37 °C, prior to applying the primary antibody.

3.2 First Immunofluorescence

1. Apply primary antibodies: Goat anti-Oct3/4, and rabbit anti-Sox17 (both Santa Cruz Biotechnology), both at the concentration of 2 μg/ml in Blocking Solution (0.6 ml) overnight at 4 °C. Note, the primary and secondary antibodies used together with the following anti-MeC antibody set may vary in their concentration (see Note 4).

2. Wash cells four times (5 min each) with Wash Buffer A (2 ml) at room temperature.

3. Apply secondary antibodies: Chicken anti-rabbit IgG (H+L)-Alexa 647 (Life Technologies A21443) and donkey anti-goat IgG (H+L)-Alexa 568 (Life Technologies A11057), both at the concentration of 5 μg/ml in Blocking Solution (0.6 ml) for 1 h at 37 °C (see Note 2). Use polyclonal antibodies that are highly cross-absorbed against several species (see Note 5).

4. Wash cells four times with Wash Buffer A (2 ml for 5 min each) at room temperature. Then, immediately proceed with treatment of cells after first antibody set.

3.3 Acid Treatment

1. Fix cells (for a second time) in Fixation Solution (1 ml) for 15 min at room temperature, then wash cells three times with Wash Buffer B (2 ml for 5 min each) at room temperature. This step will highly reduce the possibility of the first antibody set being erased through the following HCl treatment (see Note 6).

2. Depurinate cells with 2 N HCl Solution (1 ml) for 40 min at room temperature, then wash cells three times with Wash Buffer B (2 ml for 5 min each) at room temperature. Time and concentration of HCl are optimized and can be used with a variety of different cell types (Fig. 2). Deviating from this combination may result in either weaker MeC signals or obscuring of DAPI pattern and/or patterns of protein target(s) (see Note 7).

3.4 Second Immunofluorescence

1. Block cells with Blocking Solution (1 ml) for 30 min to 1 h at 37 °C (before applying the primary antibody).

2. Add primary antibody: Mouse anti-MeC (clone 33D3 mAb) from Aviva Systems Biology (Cat. No. AMM99021) or equivalent at the concentration of 1 μg/ml in blocking solution (0.6 ml) for 1 h at 37 °C (see Note 8).

3. Wash cells four times for 5 min each with Wash Buffer A (2 ml) at room temperature.

4. Add secondary antibody: Donkey anti-mouse IgG (H+L)-Alexa 488 (Life Technologies A21202) at the concentration of 5 μg/ml in blocking solution (0.6 ml) for 1 h at 37 °C (see Note 9).

5. Wash cells four times with Wash Buffer A and once with Wash Buffer B (2 ml for 5 min each) at room temperature.

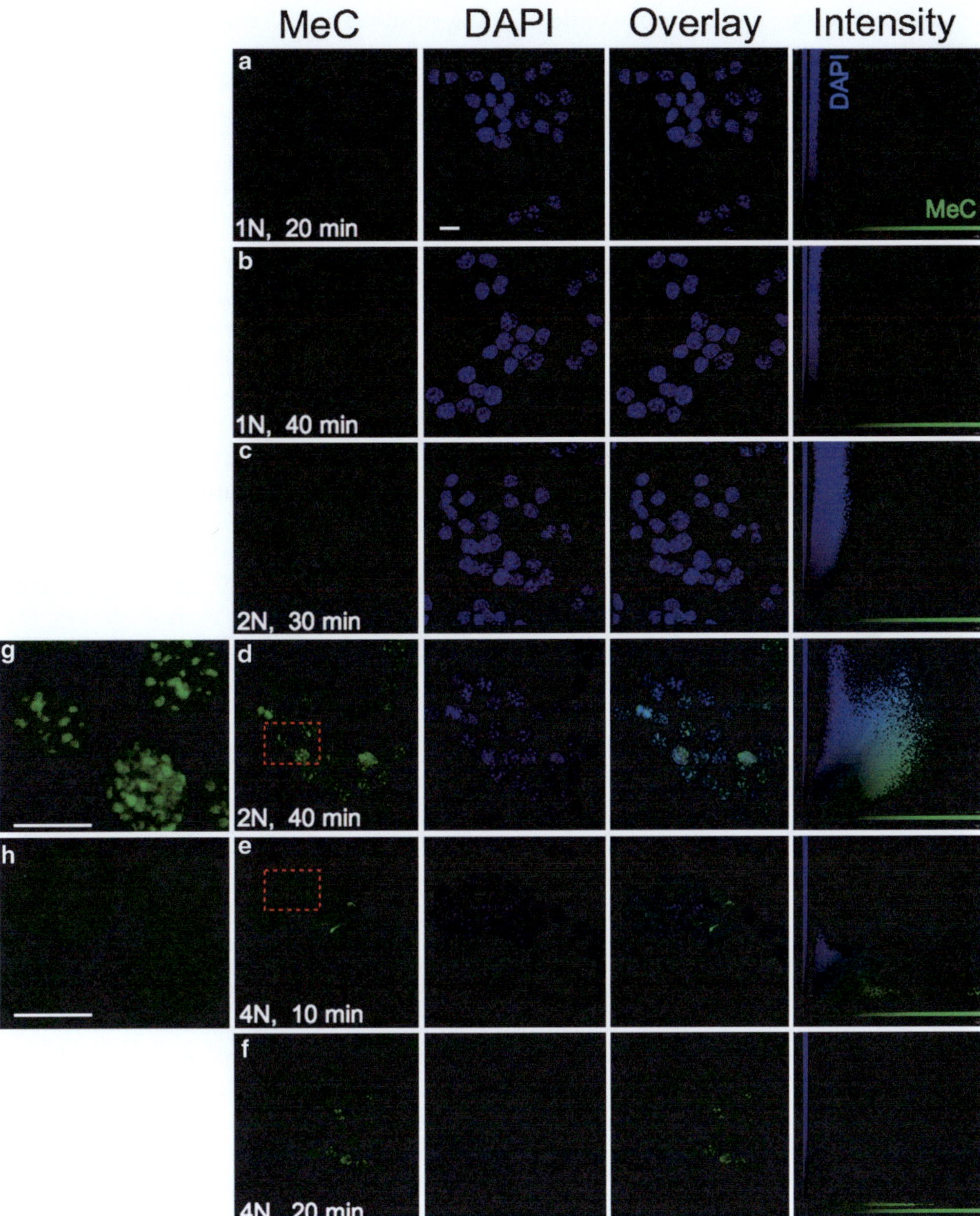

Fig. 2 Two-color imaging of differentiating mouse cells: Maximum intensity projection (MIP) of 2D confocal images taken from a cluster of fixed cells that were processed with different concentrations of HCl and different incubation times for depurination/denaturation of the double-stranded cellular DNA (rows **a–f**): Nuclear patterns of MeC (*false-colored green*) and gDNA delineated by DAPI (*false-colored blue*). The *left* column displays the distribution of the two nuclear signals as 2D intensity scatter plots. Satisfying delineation of the two classes of DNA was obtained by applying 2 N HCl for 40 min (**d**), as judged by intensity distribution and nuclear patterns, with partial colocalization of MeC and DAPI typically in larger nuclear clusters, known as chromocenters (magnified cells in **g**). The overlay of *green* (MeC) and *blue* (DAPI) appears turquoise. Using lower acid concentrations results in insufficient unmasking and detection of MeC sites (very low intensity in the *green channel*) (**a–c**). Higher acid concentrations led to a rapid signal reduction in both channels as a consequence of structural distortion (magnified cells in **h**) by excessive DNA denaturation. The size of all scale bars is 10 μm

3.5 Counterstaining and Mounting

6. Replace the washing solution with 1 ml of DAPI Counterstain Working Solution and incubate for 15 min at room temperature. For better intercalation into DNA, let DAPI solution (kept at 4 °C) reach quasi room temperature before application.

7. Carefully remove coverslip from well and transfer onto filter paper (Whatman™). Let coverslip dry at room temperature (or for acceleration in oven at 37 °C).

8. Transfer ample (7–10 µl) of ProLong Gold™ (Cat. No. P36930, Life Technologies) embedding solution onto microscopic slide. Then carefully put coverslip upside-down onto the mounting droplet. Avoid air bubbles as these change the optical properties and interfere with high-resolution imaging of entrapped cells with oil or glycerol immersion objectives. Specimen needs to be kept at room temperature for a minimum of 6 h (protected from light) to ensure hardening of embedding resin before fluorescence microscopy.

3.6 Image Acquisition and Visualization

Specimens can be analyzed by confocal laser scanning microscopy using a variety of microscopes from different manufacturers. Microscopes need to be equipped with appropriate laser lines such as a multiline argon laser (465, 488, 514 nm) and HeNe-lasers (540, 590, 633 nm). Also, a white laser provided by the TCS SP5 X Supercontinuum microscope (Leica Microsystems), can be used: The system provides full freedom and flexibility in excitation and emission, within the continuous range of 470–670 nm—in 1 nm increments. In both cases an additional 405 nm diode laser line for excitation of DAPI fluorescence will be necessary. The following steps are universal and should be consistently performed for serial image sampling that creates images at the conventional confocal resolution limit.

1. Collect optical sections at increments of 200–300 nm (Nyquist sampling density) with a Plan-Apo 63× 1.4 oil immersion lens or a Plan-Apo 63× 1.3 glycerol immersion lens, and pinhole size 1.0 airy unit. Image size should be chosen either 1,024 × 1,024 or 2,048 × 2,048 (if available) with a respective voxel size of approximately 120 nm × 120 nm × 250 nm (x-, y-, and z-axis), with a resolution of 8–12 bits per pixel in all channels (see Note 10).

2. The microscope/scanner output file can be converted to a series of TIFF images using a variety of softwares, including the freeware ImageJ (developed at the US National Institutes of Health).

3. TIFF-images are compatible with diverse commercially available and custom-made machine learning algorithms for 3D visualization of delineated cellular targets (MeC, global DNA, and protein markers), as well as topological analysis and image cytometry (imaging-based cytomics).

4 Notes

1. An appropriate fixation protocol should prevent antigen leakage, maintain cellular structure, and allow for permeabilization and efficient antigen retrieval. It is recommended to avoid organic solvents such as alcohols and ketones, which typically remove lipids, and thus dehydrate the cells and distort the subcellular structure, especially the spatial distribution of interphase chromatin. Paraformaldehyde, on the other hand crosslinks cellular entities via free amino groups, and thereby creates a meshwork that can hold the cells' structure in its authentic conformation. However PFA, usually applied as 1–4 %, reduces membrane permeability, therefore calling for an additional permeabilization step. Thus, fixation with PFA is a compromise between a better preservation of cellular structure and antigen accessibility. Therefore fixation time should be restricted to the minimum required for structure preservation to avoid possible epitope masking. In our experience, the fixation of cells and tissues with 4 % PFA for approximately 15–45 min and the subsequent permeabilization with 0.5 % saponin/0.5 % triton X-100 for 20 min at room temperature led to satisfactory results for immunofluorescence detection of MeC, gDNA, and proteins, respectively (2, 27, 30, 31).

2. Avoid dated Permeabilization Solution (slightly opaque) that can cause weak MeC signals in cells not linked to demethylation. Check solution before application: Generally, solution (kept at 2–8 °C) should not be used beyond 2 weeks of storage.

3. Even though hydrolysis is known to also disintegrate cellular RNA-species, an RNase A treatment step is integrated in the protocol to avoid any residual transfer RNA (tRNA) molecules that have escaped complete digestion and removal from the cells for the following reason: tRNAs can contain methylated cytosine. Although tRNAs are not expected within the mammalian nucleus, this step should be used for eliminating cytosolic tRNA species in favor of a better demarcation and contrasting of the nucleus against the cytosol (in the MeC image channel). Furthermore, in cases such as HIV infected cells, it has been shown that defective tRNA is shuttled into the nucleus (32).

4. Check compatibility of antibody with other primary antibodies used in the assay. Avoid using an antibody raised in mouse, as the downstream applied anti-MeC antibody is a monoclonal mouse antibody.

5. Always perform negative control experiments by applying combination of secondary antibodies without their respective primary antibodies. If any signal above the average background noise—defined as two standard deviations above the average

signal outside the nucleus or the cell—is detected, then nonspecific staining of secondary antibody due to lack of antibody quality has occurred. Use a secondary antibody from another vendor, or even species. Sometimes higher signal background can be experienced in specimens due to nonspecific staining of coverslip coating (such as poly-L-lysine) for the attachment of naturally nonadherent cells. In that case try blocking with solutions that have a BSA portion of higher than 3 % w/v or use serum of the secondary antibody species.

6. It is recommended to apply immunostaining of protein markers before HCl treatment, followed by a secondary fixation step with PFA that crosslinks antibodies, antigen, and conjugated fluorophores in order to protect the complex from acid-induced degradation. Nevertheless, in case if the protein signal (fluorescence) is weaker than previously experienced under equivalent conditions, this may be due to weak crosslinking of cellular proteins. In this case, try longer initial fixation times of up to 45 min. Note, longer fixation times with paraformaldehyde, however, may reduce membrane permeability (as mentioned above), and in turn may require more rigorous MeC antigen retrieval, i.e. eventually heat denaturation in addition to acid treatment.

7. The developed protocol employs HCl treatment at a dose and duration that allows for the visualization of MeC sites without a detectable (by high-resolution confocal microscopy) adverse effect on DAPI staining. At this combination we did not observe any detectable differences in gDNA staining patterns (as nuclear DAPI intensity distributions), in comparison to cells that were not treated with HCl and/or not stained for MeC. Treatment of cells with lower doses of HCl or at a shorter duration resulted in equivalent DAPI intensity patterns but decreased MeC intensities, possibly due to insufficient unmasking of MeC sites. Depurination with higher acid concentrations had a double effect: (1) DAPI intensity was gradually reduced, and (2) both, the DAPI and the MeC patterns appeared fuzzier in confocal microscopic images. Both phenomena can be interpreted as a sign of DNA decomposition in response to excessive denaturation. Additionally, acid treatment can alter protein epitopes and subsequently impair recognition and the immunodetection of protein targets. To verify the protocol compatibility with covisualization of targeted proteins, control experiments are recommended, in which the distribution of targeted proteins is compared between HCl treated and untreated cells.

8. As the anti-MeC antibody is a key reagent in DNA methylation phenotyping, it is necessary to assure specificity and sensitivity of the antibody. This can be performed with a dot blot assay (north-western hybridization): Practically, the antibody is

hybridized at various concentrations to an immobilized array of template DNA with differential numbers of methylated cytosine residues, followed by colorimetric detection (33). A modification of this technology utilizes a glass slide-based microarray of immobilized synthetic oligonucleotides as probes in combination with fluorescence detection using the same combination of primary anti-MeC antibody and secondary fluorophore-conjugated antibody as the *in situ* assay (34).

9. If the negative control (no anti-MeC antibody) shows nonspecific signal in the respective channel, this could be highly possibly through cross-reaction of the secondary antibodies (mentioned above). First, try to reduce the concentration of the anti-MeC antibody. If that does not help in significantly reducing nonspecific staining, use a combination of same type of secondary antibodies from other vendors or even other species.

10. To avoid bleed-through of channels, the imaging of individuals channel can be acquired sequentially.

Acknowledgements

Kolja Wawrowsky's (Cedars-Sinai Medical Center) consultation on confocal microscopy is thankfully appreciated. This work was supported by DoD award W81XWH-10-1-0939 and internal seed grants from the Department of Surgery at CSMC.

References

1. Wang D, Bodovitz S (2010) Single cell analysis: the new frontier in 'omics'. Trends Biotechnol 28:281–290

2. Tajbakhsh J, Gertych A, Fagg WS, Hatada S et al (2011) Early in vitro differentiation of mouse definitive endoderm is not correlated with progressive maturation of nuclear DNA methylation patterns. PLoS One 6:e21861

3. Erlanger BF, Beiser SM (1964) Antibodies specific for ribonucleosides and ribonucleotides and their reaction with DNA. Proc Natl Acad Sci U S A 52:68–74

4. Reynaud C, Bruno C, Boullanger P et al (1992) Monitoring of urinary excretion of modified nucleosides in cancer patients using a set of six monoclonal antibodies. Cancer Lett 63:81

5. Schreck RR, Erlanger BF, Miller OJ (1974) The use of antinucleoside antibodies to probe the organization of chromosomes denatured by ultraviolet irradiation. Exp Cell Res 88:31–39

6. Miller OJ, Schnedl W, Allen J et al (1974) 5-Methylcytosine localised in mammalian constitutive heterochromatin. Nature 251:636–637

7. Miniou P, Jeanpierre M, Blanquet V et al (1994) Abnormal methylation pattern in constitutive and facultative (X inactive chromosome) heterochromatin of ICF patients. Hum Mol Genet 3:2093–2102

8. de Capoa A, Menendez F, Poggesi I et al (1996) Cytological evidence for 5-azacytidine-induced demethylation of the heterochromatic regions of human chromosomes. Chromosome Res 4:271–276

9. Rougier N, Bourc'his D, Gomes DM et al (1988) Chromosome methylation patterns during mammalian preimplantation development. Genes Dev 12:2108–2113

10. Bensaada M, Kiefer H, Tachdjian G et al (1998) Altered patterns of DNA methylation on chromosomes from leukemia cell lines: identification of 5-methylcytosines by indirect immunodetection. Cancer Genet Cytogenet 103:101–109

11. Montpellier C, Burgeois CA, Kokalj-Vokac N et al (1994) Detection of methylcytosine-rich heterochromatin on banded chromosomes. Application to cells with various status of

DNA methylation. Cancer Genet Cytogenet 78:87–93

12. Barbin A, Montpellier C, Kokalj-Vokac N et al (1994) New sites of methylcytosine-rich DNA detected on metaphase chromosomes. Hum Genet 94:684–692

13. de Capoa A, Grappelli C, Febbo FR et al (1999) Methylation levels of normal and chronic lymphocytic leukemia B lymphocytes: computer-assisted quantitative analysis of anti-5-methylcytosine antibody binding to individual nuclei. Cytometry 36:157–159

14. de Capoa A, Febbo FR, Giovannelli F et al (1999) Reduced levels of poly(ADP-ribosyl) ation result in chromatin compaction and hypermethylation as shown by cell-by-cell computer-assisted quantitative analysis. FASEB J 13:89–93

15. de Capoa A, Di Leandro M, Grappelli C et al (1998) Computer-assisted analysis of methylation status of individual interphase nuclei in human cultured cells. Cytometry 31:85–92

16. Piyathilake CJ, Johanning GL, Frost AR et al (2000) Immunohistochemical evaluation of global DNA methylation: comparison with in vitro radiolabeled methyl incorporation assay. Biotech Histochem 75:251–258

17. Piyathilake CJ, Frost AR, Bell WC et al (2001) Altered global methylation of DNA: an epigenetic difference in susceptibility for lung cancer is associated with its progression. Hum Pathol 32:856–862

18. Soares J, Pinto AE, Cunha CV et al (1999) Global DNA hypomethylation in breast carcinoma: correlation with prognostic factors and tumor progression. Cancer 85:112–118

19. Dolbeare F, Gratzner H, Pallavicini MG et al (1983) Flowcytometric measurement of total DNA content and incorporated bromodeoxyuridine. Proc Natl Acad Sci U S A 80:5573–5577

20. Mayer W, Niveleau A, Walter J et al (2000) Demethylation of the zygotic paternal genome. Nature 403:501–502

21. Barton SC, Arney KL, Shi W et al (2001) Genome-wide methylation patterns in normal and uniparental early mouse embryos. Hum Mol Genet 10:2983–2987

22. Santos F, Hendrich B, Reik W et al (2002) Dynamic reprogramming of DNA methylation in the early mouse embryo. Dev Biol 241:172–182

23. Santos F, Zakhartchenko V, Stojkovic M et al (2003) Epigenetic marking correlates with developmental potential in cloned bovine preimplantation embryos. Curr Biol 13:1116–1121

24. Sasaki K, Adachi S, Yamamoto T et al (1988) Effects of denaturation with HCl on the immunological staining of bromodeoxyuridine incorporated into DNA. Cytometry 9:93–96

25. Kennedy BK, Barbie DA, Classon M et al (2000) Nuclear organization of DNA replication in primary mammalian cells. Genes Dev 14:2855–2868

26. Grskovic M, Javaherian A, Strulovici B et al (2011) Induced pluripotent stem cells—opportunities for disease modelling and drug discovery. Nat Rev Drug Discov 10:915–929

27. Tajbakhsh J, Wawrowsky KA, Gertych A et al (2008) Characterization of tumor cells and stem cells by differential nuclear methylation imaging. In: Farkas DL, Nicolau DV, Leif RC (eds) Imaging, manipulation, and analysis of biomolecules, cells, and tissues VI. Proceedings, vol 6859, 68590F

28. Tajbakhsh J, Luz H, Bornfleth H et al (2000) Spatial distribution of GC- and AT-rich DNA sequences within human chromosome territories. Exp Cell Res 255:229–237

29. Scheuermann MO, Tajbakhsh J, Kurz A et al (2004) Topology of genes and nontranscribed sequences in human interphase nuclei. Exp Cell Res 301:266–279

30. Gertych A, Wawrowsky KA, Lindsley EH et al (2009) Automated quantification of DNA demethylation effects in cells via 3D mapping of nuclear signatures and population homogeneity assessment. Cytometry A 75:569–583

31. Gertych A, Farkas DL, Tajbakhsh J (2010) Measuring topology of low-intensity DNA methylation sites for high-throughput assessment of epigenetic drug-induced effects in cancer cells. Exp Cell Res 316:3150–3160

32. Zaitseva L, Myers R, Fassati A (2006) tRNAs promote nuclear import of HIV-1 intracellular reverse transcription complexes. PLoS Biol 4: e332

33. Tao L, Wang W, Kramer PM (2004) Modulation of DNA hypomethylation as a surrogate endpoint biomarker for chemoprevention of colon cancer. Mol Carcinog 39:79–84

34. Tajbakhsh J, Gertych A (2012) 3-D Quantitative DNA methylation imaging for chromatin texture analysis in pharmacoepigenomics and toxicoepigenomics. In: Appasani K (ed) From chromatin biology to therapeutics. Cambridge University Press, Cambridge, UK

Methods Molecular Biology (2013) 1052: 89–99
DOI 10.1007/7651_2013_26
© Springer Science+Business Media New York 2013
Published online: 3 May 2013

Noninvasive Imaging of Myocardial Blood Flow Recovery in Response to Stem Cell Intervention

HuaLei Zhang and Rong Zhou

Abstract

The recovery of myocardial blood flow is a major indicator of the effectiveness of cell-based therapies for ischemic heart diseases including myocardial infarction. Blood flow (also called perfusion) of the heart muscle can be noninvasively measured via imaging methods such as ultrasound, positron emission tomography (PET), or magnetic resonance imaging (MRI). Here, we describe an MRI technique, namely, spin labeling, to measure the volumetric blood flow (mL/min/g) in the heart. Specifically, we demonstrate how impaired blood flow in the infarcted region of the heart was recovered transiently (≥ 2 weeks) after the injection of endothelial progenitor cells.

Keywords: Stem cell, Magnetic resonance imaging, Blood flow, HUVEC

1   Introduction

Magnetic resonance imaging (MRI) provides high resolution and excellent soft tissue contrast and, therefore, has become the gold standard to evaluate global cardiac function parameters, such as ejection fraction, particularly in infarcted hearts undergoing cell-based therapies (1–5). In addition, the versatility and potential of MRI to assess infarct size, regional contractile function, perfusion, and metabolism in a single examination have been well recognized (6). These advanced MRI techniques would facilitate the understanding of mechanisms underlying therapeutic interventions. In specific, one of the myocardial function indices—blood flow (perfusion)—has been estimated commonly in two ways by MRI: (1) injection of exogenous contrast agent (such as Gd-DTPA, Magnevist®) to capture the first-pass dynamic contrast enhancement (DCE) or dynamic susceptibility contrast (DSC) features and (2) labeling the arterial blood with radiofrequency pulses as endogenous tracer to assess the MR signal change as arterial blood perfuses into the heart tissue, a technique formally called spin labeling.

The first method entails elaborate modeling of the dynamic information to obtain quantitative measurements (7). It has been commonly used in humans due to the short acquisition time (40–50 heartbeats) and well-enhanced images; however, the demand of high spatial and temporal resolutions limits its use in small animal hearts, given their fast rates (400–600 beats per minute, bpm, compared to 60–80 bpm in humans) and small sizes (about 1/200th of human hearts in volume). In comparison, the second method, spin labeling, is able to overcome these challenges and has been employed extensively to measure blood flow in the heart of small rodent models (8–10). Furthermore, the method allows repeated measurements since it is not necessary to wait for clearance of the exogenous contrast agent as in DCE–MRI; the spin labeling technique is also more suitable for scanners operating at high magnetic field (>3 T) (11), where small animal models are usually studied.

Here we demonstrate how pixel-wise mapping of myocardial blood flow (MBF) is obtained by the spin labeling technique, and how regional blood flow is depressed in the infarcted heart and how it recovers in response to the injection of endothelial cells in the heart. The flowchart in Fig. 1 illustrates a typical work flow of cell-based experimental therapies and MBF measurements of infarcted hearts pre- and posttreatment.

2   Materials

2.1 Endothelial Cells and Culture Medium

1. Human umbilical vein endothelial cells (HUVEC, ATCC, Manassas, VA), used within a total of 6–10 passages.

2. Growth factor-reduced Matrigel (Collaborative Biomedical, Bedford, MA).

2.2 Antibodies Used to Visualize Engrafted Cells and Capillaries

1. Rabbit polyclonal anti-von Willebrand Factor (vWF) antibody (Sigma, St. Louis, MO) combined with FITC-conjugated goat anti-rabbit secondary antibody (Santa Cruz Biotechnology, Santa Cruz, CA).

2. Mouse monoclonal anti-human CD34 antibody (Abcam, Cambridge, MA) combined with Cy3-conjugated goat anti-mouse secondary antibody (Abcam, Cambridge, MA).

2.3 Hardware Requirement in Spin Labeling-MRI Acquisition

1. MR scanner: The parameters and results presented here are based on 4.7 T magnetic field with a maximum gradient strength of 25 G/cm (see Note 1).

2. Coil specs: For cardiac imaging, a combination of volume transmit and surface receive coils is recommended, as illustrated in Fig. 2. The volume coil should have the animal body closely fill up the space for a good performance of the coil, and cover the body length for proper global inversion of

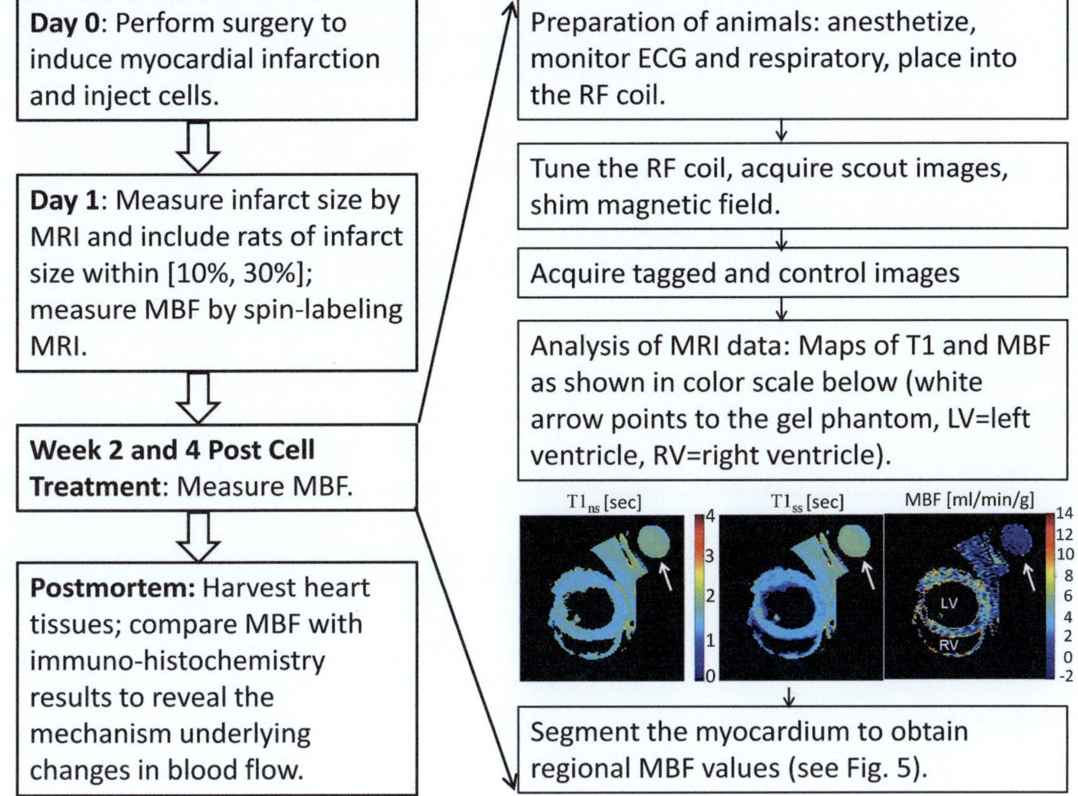

Fig. 1 A flowchart demonstrating the evaluation of longitudinal MBF changes in the infarcted heart subjected to cell-based therapies

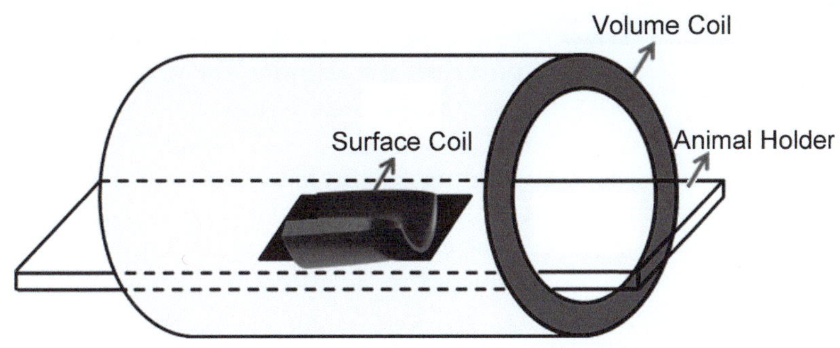

Fig. 2 The coil set combining a volume coil for radio frequency (RF) transmission and a surface coil for signal receiving. The animal holder had a rectangular opening (*black area*) to embed the surface coil on the board

the blood. The surface coil has an arc curvature to fit the animal chest. A decoupling mechanism is required to switch off the surface-receive coil during the transmission of radiofrequency pulses by the volume coil. In our studies, a volume-transmit coil with internal diameter of 71 mm and a surface-receive coil with a curve arc corresponding to 120° of a full circle of 15-mm diameter (InsightMRI, Worcester, Mass) were used for cardiac imaging of rats (see Note 2).

3   Methods

3.1 Animal Model of Myocardial Infarction and Intramyocardial Injection of Endothelial Cells

1. Myocardial infarction (MI) was induced in athymic nu/nu rats (6–8 weeks old; Frederick Cancer Center, Frederick, MD) by permanent ligation of the left anterior descending (LAD) coronary artery.

2. During the MI surgical session, 5 million HUVECs in 100 μL Matrigel, or Matrigel alone (vehicle), were injected in the border close to the blanched (infarcted) area (12) (see Note 3).

3. Initial infarct size (as a percentage of left ventricle myocardial volume) was assessed by MRI (see Note 4) 24–48 h after MI surgery and served as an exclusion criterion (see Note 5) for animals under further study.

3.2 Electrocardiogram and Respiratory Gating Setup

1. Animals were maintained under isoflurane (1.5 % v/v with oxygen) via a nose cone and placed prone in the RF coil.

2. Electrocardiogram (ECG) signal was obtained by attaching needle-type platinum leads in the right front paw and left hind paw.

3. Respiratory waveform was obtained by placing a pneumatic pillow underneath the abdomen of the rat.

4. Core temperature was maintained at 36.5 ± 0.2 °C by directing warm air into the magnet bore.

5. Trigger signals of ECG and respiration were connected to separate ports on the scanner for proper synchronization of MRI data acquisition (see Note 6).

3.3 MR Image Acquisition

On-slice spin tagging scheme is illustrated in Fig. 3, with non-slice-selective (Fig. 3a) or slice-selective (Fig. 3b) inversion of magnetization to produce corresponding control or tagged images.

A series of images are acquired during inversion recovery for T1 mapping (Fig. 4) based on a T1-by-multiple-readout-pulses (TOMROP) sequence:

1. A 180° inversion pulse is applied upon the first ECG trigger coincident with the beginning of a respiration plateau.

2. A set of images are acquired in a segmented k-space scheme every 1–2 heartbeats (see yellow boxes in Fig. 4) for approximately 7 s; due to the rapid heart rate, each segment contains 1–2 phase encoding k-space lines in our study.

3. A delay of 4 s ensures a full recovery of the longitudinal magnetization (see Note 7), before the next inversion pulse is applied.

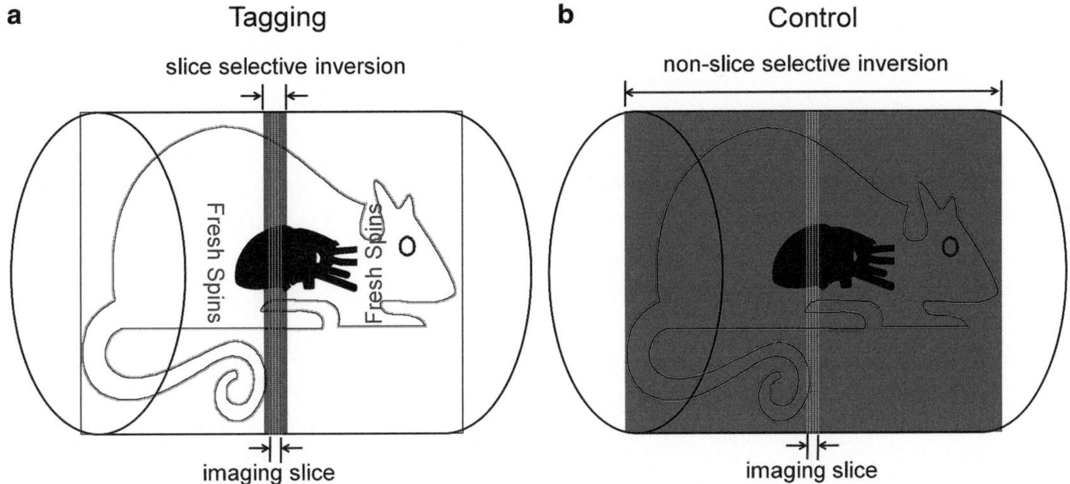

Fig. 3 The flow-sensitive alternating inversion-recovery (FAIR) pulse sequence generates tagged and control images. The tagged images are produced after applying a slice-selective 180° inversion pulse to the imaging slice; the plane of this slice is marked by the thin *gray* slab through mid left ventricle as shown in (**a**). Since the tagged spins shown in (**a**) are not inverted, their entry into the imaging slice as the result of blood flow will modulate the T1 relaxation of the imaging slice. See Note 10 for determining the optimal ratio between the thickness of inversion slice versus imaging slice in (**a**). The control images are produced at the same plane as marked in (**a**) but after applying a non-slice-selective inversion pulse to the entire sensitive volume of the RF coil as shown by the *gray* area in (**b**)

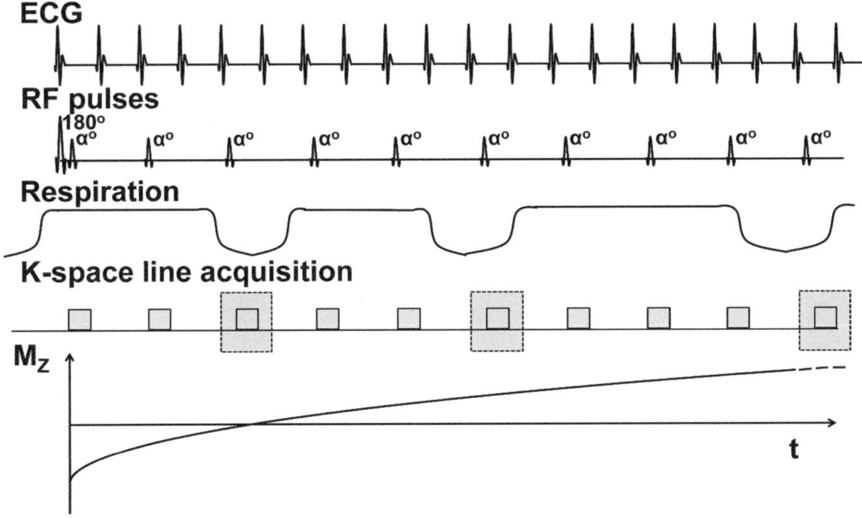

Fig. 4 The acquisition scheme of TOMROP sequence with physiological signal monitoring

4. The loop of steps 1–3 is repeated until the k-space is filled out. Maps of $T1_{ss}$ or $T1_{ns}$ are produced, respectively, corresponding to the slice-selective (Fig. 3a) and non-slice-selective (Fig. 3b) inversion of magnetization.

5. If more signal averages are needed to increase the signal-to-noise ratio (SNR), alternating the acquisition of $T1_{ns}$ and $T1_{ss}$ maps is recommended to reduce the influence of physiological fluctuations on the maps.

6. During the acquisition, recording the respiratory waveform and RF pulse timings (see Note 8) facilitates to identify motion-corrupted k-space lines/images (see dotted boxes in Fig. 4) and retrieve heart rate statistics (see Note 9) for improving accuracy in T1 fitting.

3.4 MR Parameter Adjustments

1. The imaging slice thickness: A 3 mm mid-ventricular slice is chosen for trade-offs between SNR and through-plane partial volume effect (the left ventricle typically spans 12 mm from the base to the apex of the heart for a 200 g rat).

2. Multiplication factor of the inversion slice thickness: To better match slice profiles of the hyperbolic shaped inversion pulse and SINC-shaped excitation pulse, the inversion pulse thickness was calibrated to be 2.5 times of imaging slice thickness (see Note 10).

3. The number of image points captured during inversion recovery for T1 estimation: It was calculated as a rounding integer of 7 s divided by the gap time between excitation pulses, which was planned based on the heart rate in each scan.

3.5 Image Analysis

1. The intensity of any pixel within area of interest M_n (n = 1...N, N is number of image points) can be modeled as a function of equilibrium magnetization M_{eq}, excitation pulse α, and longitudinal relaxation T1 (13, 14).

2. A 3-parameter least square fitting was performed to calculate pixel-wise T1 values under non-slice-selective and slice-selective inversions, designated as $T1_{ns}$ and $T1_{ss}$, respectively.

3. The pixel-wise blood flow MBF was quantified (15) based on the equation below:

$$\frac{\text{MBF}}{\lambda} = \frac{T1_{ns}}{T1_b} \cdot \left(\frac{1}{T1_{ss}} - \frac{1}{T1_{ns}} \right) \tag{1}$$

where λ is the tissue–blood partition coefficient of water and set to 0.83 mL/g for rat myocardium, and $T1_b$ is the blood T1 under global inversion.

4. The gel phantom is included in the T1 mapping and a blood flow measurement is accepted if the phantom has a mean flow value within ± 0.5 mL/min/g; otherwise the data is considered invalid (due to unstable heart beats and/or respiratory cycles or other reasons) and should be reacquired if possible.

5. To obtain regional blood flow, the left ventricle myocardium in short axis view was segmented into Infarcted (I)-Border (B)-Remote (R), where infarcted region was defined on the LGE image (see Note 3) acquired at the same position as spin labeling imaging slice, border region was defined as two sectors of 30° on each side of the infarcted segment, and the remote region encompassed the remaining myocardium.

3.6 Immunohisto-logical Analyses

Microvasculature density (MVD) was estimated at 2 and 4 weeks based on immunohistochemical staining of the sections as we described previously (12).

1. Under 2× objective lens, each section was segmented into the I-B-R region: The infarcted region was identified on the adjacent hematoxylin and eosin section; the border and remote regions were defined in the same way as on MR images.

2. Under 10× objective lens, at least 6 field-of-views (each covering 1.1 mm^2) were captured including 3 in the remote, 2 in the border, and 1–3 in the infarcted region (depending on the infarct size). Clustered cells or continuous branching structures with positive vWF staining were counted as 1 capillary.

3.7 Result Interpretation

1. The regional MBF was impaired in the infarcted region compared to the remote region at 1 day post MI (inset of Fig. 5a).

2. At 2 weeks post injection of endothelial cells (EC), MBF averaged over the entire slice was significantly higher in the EC group than in the vehicle group ($P = 0.0380$, Fig. 5a).

3. When the analysis was refined to a specific region or time point, a significant treatment effect in the infarcted region at 2 weeks ($P = 0.0086$) and in the remote region at 4 weeks ($P = 0.0277$) was obtained. Representative MBF maps at 2 weeks indeed showed higher blood flow in the infarcted segment in the EC-treated heart compared to the vehicle-treated one while both had a similar infarct size (Fig. 5b–e).

4. In the infarcted region more capillaries in EC-treated hearts were revealed by vWF immunostaining (Fig. 5f–g); quantitative analysis confirmed a significantly higher MVD in the infarcted region ($P = 0.0105$ in EC versus vehicle group, Fig. 5j). Furthermore, double staining for vWF and human CD34 demonstrated the incorporation of ECs into capillaries in the infarcted (Fig. 5h) and border region (Fig. 5i).

These data provide convincing evidence that EC engraftment enhanced new vessel formation, leading to improved perfusion in the infarcted region detected by spin labeling MRI technique.

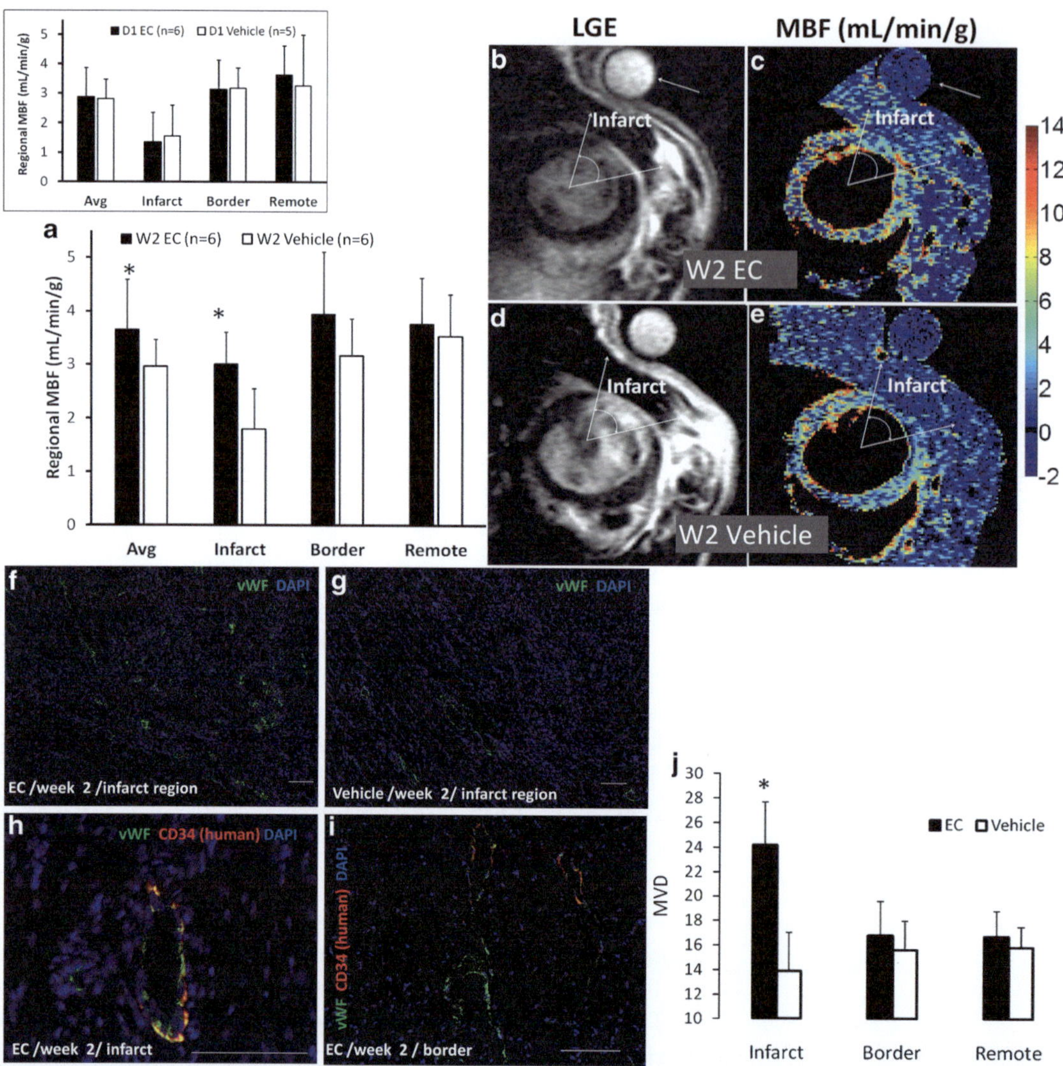

Fig. 5 Regional myocardial blood flow estimated by in vivo MRI and microvasculature density (MVD) at 2 weeks after MI. Slice average and regional blood flow at 1 day (*inset* of **a**) and 2 weeks after MI (**a**) are shown. The MBF maps and corresponding late gadolinium enhancement (LGE) images are shown for a representative heart from the endothelial cell (EC) group (**b** and **c**, *arrows* pointing to the gel phantom) and the vehicle group (**d** and **e**). Immunostaining for von Willebrand Factor (vWF) from an EC-treated (**f**) or vehicle-treated heart (**g**) is shown. Double immunostaining of CD34, which reacts only to human tissue, and vWF antibody, which reacts to rat and human antigen, revealed incorporation of ECs into capillaries in the infarcted (**h**) and border (**i**) regions. Scale bars = 100 μm. MVD = the number of capillaries per field of 1.1 mm^2. (*Asterisk*) Statistically significant comparing the EC versus the vehicle group (**a** and **j**). (From Zhang et al. (12), reprinted with permission)

4   Notes

1. At higher magnetic fields, it increases SNR and tissue longitudinal relaxation time (T1). The latter effect enhances the difference of T1 in tagged and control conditions, together with improved SNR, leading to more reliable perfusion measurements, but meanwhile prolongs the data acquisition time for accurate T1 mapping.

2. Regarding the animal holder, we constructed a piece of Teflon plastic plate with a rectangular opening to snug the surface coil, and a separable nose cone to adjust positions for animals of varied sizes.

3. If the cells are injected during a separate surgical session after the myocardial infarction surgery, the lateral intercostal space can be used as the entry site to avoid reentry from the sternum (16). This strategy avoids excessive bleeding from dissection of scar tissue formed after the first surgery and more importantly reduces surgery time as well as animal mortality.

4. Late gadolinium enhancement (LGE) is an MRI technique to measure the infarction size (17). After intravenous injection of 0.3–0.6 mmol/kg Gd-DTPA, a delay of 10–20 min is needed so that in normal myocardium the contrast agent washes out. T1-weighted images are then acquired, in which the infarcted area is enhanced due to delayed washout of Gd-DTPA contrast agent. By delineating infarction area in LGE images, the infarction size can be estimated overall as the percentage of LV volume.

5. Rats having an infarct size within the range of (10 %, 30 %) were included for subsequent studies outlined in the flowchart (Fig. 1), while others were excluded. We have consistently applied such exclusion procedure in our previous studies (12, 16, 18–20); it ensures a relatively uniform infarct size in study groups and reduces bias introduced by very small or large infarct sizes because the degree of functional recovery is dependent on engrafted cells and the initial infarct size as well.

6. We recommend to apply the inversion RF pulse after a double trigger, i.e., by first detecting the front of respiratory plateau followed by detection of an ECG before application of the RF pulse.

7. The time lengths (7-s acquisition followed by 4-s delay) are customized based on an estimated rat myocardium T1 of 1.3–1.6 s at 4.7 T and acquisition parameters including the excitation flip angle and gap time between excitation pulses. The goal here is for the longest T1 component to approach the steady-state condition by the end of the excitation pulse train, and to fully relax to equilibrium magnetization by the end of the delay.

8. Notes 8 and 9 are refinements of the spin labeling technique discussed in the main text. While the first image following the inversion pulse is triggered by both respiratory and ECG, the remaining images are triggered by ECG only and hence acquired at all possible respiratory phases. We would like to filter out ones that are acquired during the inspiration phase to improve the accuracy of the T1 fitting. To achieve this, the first step is to record the timing of image acquisitions relative to the respiratory waveform—discussed here. The second step is to determine which images should be excluded—discussed in Note 9. The respiratory waveform and a TTL signal synchronized with RF excitation pulses can be simultaneously recorded by a signal breakout module (SA Instrument, Stony Brook, NY) (12). The pulse sequence is programmed to turn on a TTL signal right after each ECG-triggered excitation pulse for a few milliseconds (if the recording rate is 200 Hz, the TTL is on for at least 5 ms).

9. The gap time between ECG-triggered excitation pulses is necessary for T1 fitting. With the recording of RF pulse timings, the average cardiac cycle within each T1 map acquisition can be calculated as the weighted mean of time spacings. An apodizing function, such as a bell-shaped Hann window, puts lower weight on time spacings collected in the peripheral k-space which contributes less to the image SNR.

   The respiratory phases are categorized into inspiration and expiration. With the recording of respiratory waveform, the threshold level to classify the expiration phase can be estimated using least-square smoothing filters such as a Savitzky–Golay filter. This filter is able to accommodate the diaphragm drifting over the long scan. An image is discarded in the T1 fitting if its central k-space was collected mostly outside the quiescent expiration phase.

10. Using an agarose gel phantom (i.e., under the no-flow condition), the flow of the phantom was estimated when the ratio was set as 1, 1.5, 2, 2.5, 3 or 3.5. The smallest ratio that can achieve the "zero" flow results in the phantom is defined as the optimal ratio between the thickness of inversion slice versus imaging slice.

References

1. Lunde K, Solheim S, Aakhus S et al (2006) Intracoronary injection of mononuclear bone marrow cells in acute myocardial infarction. N Engl J Med 355(12):1199–1209

2. Assmus B, Honold J, Schachinger V et al (2006) Transcoronary transplantation of progenitor cells after myocardial infarction. N Engl J Med 355(12):1222–1232

3. Johnston PV, Sasano T, Mills K et al (2009) Engraftment, differentiation, and functional benefits of autologous cardiosphere-derived cells in porcine ischemic cardiomyopathy. Circulation 120(12):1075–1083

4. Meyer GP, Wollert KC, Lotz J et al (2006) Intracoronary bone marrow cell transfer after myocardial infarction: eighteen months'

follow-up data from the randomized, controlled BOOST (BOne marrOw transfer to enhance ST-elevation infarct regeneration) trial. Circulation 113(10):1287–1294

5. Janssens S, Dubois C, Bogaert J et al (2006) Autologous bone marrow-derived stem-cell transfer in patients with ST-segment elevation myocardial infarction: double-blind, randomised controlled trial. Lancet 367 (9505):113–121

6. Zhang H, Qiao H, Ferrari VA et al (2012) Imaging cell therapy for myocardial regeneration. Curr Cardiovasc Imaging Rep 5 (1):53–59. doi:10.1007/s12410-011-9119-z

7. Li X, Springer CS Jr, Jerosch-Herold M (2009) First-pass dynamic contrast-enhanced MRI with extravasating contrast reagent: evidence for human myocardial capillary recruitment in adenosine-induced hyperemia. NMR Biomed 22(2):148–157

8. Kober F, Iltis I, Cozzone PJ et al (2005) Myocardial blood flow mapping in mice using high-resolution spin labeling magnetic resonance imaging: influence of ketamine/xylazine and isoflurane anesthesia. Magn Reson Med 53 (3):601–606

9. Vandsburger MH, Janiczek RL, Xu Y et al (2010) Improved arterial spin labeling after myocardial infarction in mice using cardiac and respiratory gated look-locker imaging with fuzzy C-means clustering. Magn Reson Med 63(3):648–657

10. Waller C, Kahler E, Hiller KH et al (2000) Myocardial perfusion and intracapillary blood volume in rats at rest and with coronary dilatation: MR imaging in vivo with use of a spin-labeling technique. Radiology 215 (1):189–197

11. Golay X, Petersen ET (2006) Arterial spin labeling: benefits and pitfalls of high magnetic field. Neuroimaging Clin N Am 16 (2):259–268

12. Zhang H, Qiao H, Frank RS et al (2012) Spin-labeling magnetic resonance imaging detects increased myocardial blood flow after endothelial cell transplantation in the infarcted heart. Circ Cardiovasc Imaging 5(2):210–217

13. Brix G, Schad LR, Deimling M et al (1990) Fast and precise T1 imaging using a TOMROP sequence. Magn Reson Imaging 8(4):351–356

14. Pickup S, Wood AK, Kundel HL (2004) A novel method for analysis of TOROM data. J Magn Reson Imaging 19(4):508–512

15. Belle V, Kahler E, Waller C et al (1998) In vivo quantitative mapping of cardiac perfusion in rats using a noninvasive MR spin-labeling method. J Magn Reson Imaging 8 (6):1240–1245

16. Qiao H, Zhang H, Yamanaka S et al (2011) Long-term improvement in postinfarct left ventricular global and regional contractile function is mediated by embryonic stem cell-derived cardiomyocytes. Circ Cardiovasc Imaging 4(1):33–41

17. Thomas D, Dumont C, Pickup S et al (2006) T1-weighted cine FLASH is superior to IR imaging of post-infarction myocardial viability at 4.7 T. J Cardiovasc Magn Reson 8 (2):345–352

18. Zhou R, Thomas DH, Qiao H et al (2005) In vivo detection of stem cells grafted in infarcted rat myocardium. J Nucl Med 46(5):816–822

19. Qiao H, Zhang H, Zheng Y et al (2009) Embryonic stem cell grafting in normal and infarcted myocardium: serial assessment with MR imaging and PET dual detection. Radiology 250(3):821–829

20. Zhang H, Qiao H, Bakken A et al (2011) Utility of dual-modality bioluminescence and MRI in monitoring stem cell survival and impact on post myocardial infarct remodeling. Acad Radiol 18(1):3–12

Methods Molecular Biology (2013) 1052: 101–108
DOI 10.1007/7651_2013_20
© Springer Science+Business Media New York 2013
Published online: 15 May 2013

Live Imaging of Early Mouse Embryos Using Fluorescently Labeled Transgenic Mice

Takaya Abe, Shinichi Aizawa, and Toshihiko Fujimori

Abstract

Live imaging is a powerful approach to understanding the dynamic processes that occur during development. Periimplantation mouse embryos are transparent, making them suitable for live imaging. In this chapter we describe the culturing and live imaging of mouse embryos, and also introduce the reporter mouse lines for organelle labeling we recently established.

Keywords: Mouse embryo, Live imaging, Embryo culture, Time-lapse observation, Reporter mouse lines

1 Introduction

Observations of embryonic development through time-lapse imaging provide a large amount of high-quality information for understanding animal development, including early mouse development. Specific fluorescent protein probes can be introduced into mouse genome by transgenesis to help visualize developmental processes. We have produced a series of reporter mouse lines to visualize specific organelles by homologous recombination in ES cells (1). For live imaging of mouse embryonic development, it is important to select appropriate reporter transgenic mouse lines to visualize specific structures in cells, as well as to select an appropriate imaging apparatus. The culturing of early mouse embryos has been carried out by many researchers, and combining these whole embryo culture techniques with imaging techniques is essential for successful time-lapse observations of mouse embryonic development.

2 Materials

2.1 Mice and Embryos

Mice used for the live imaging should be selected from the list (Table 1). See note 1. Mouse embryos should be collected according to the standard protocol (2).

2.2 Imaging Equipment

1. Olympus incubation imaging system LCV100 equipped with a CSU10 (YOKOGAWA) and an iXon+EMCCD camera (Andor), or alternatively CV1000 (YOKOGAWA). See note 2.

2.3 Materials for Embryo Collection and Culture of Preimplantation Embryos

1. Culture and flashing medium for preimplantation embryos: KSOM (KSOM-AA MR-106-D, Millipore).
2. 35 mm glass bottom dish (P35G-1.5-10-C, MatTek).
3. Mineral oil, Light (0121-4, Fischer).
4. Fine tweezers.
5. Fine scissors.
6. Syringes (1 ml) and needles (31 G, blunt end).
7. Transfer pipette (around 200 μm diameter) with a mouthpiece.

Table 1
List of Rosa26 reporter mouse lines

Subcellular location	Fluorescent fusion protein	Line name	CDB Acc. no
Nucleus	H2B-EGFP	R26R/R26-H2B-EGFP	CDB0203K/0238K
	H2B-mCherry	R26R/R26-H2B-mCherry	CDB024K/0239K
Mitochondria	Mito-EGFP	R26R/R26-Mito-EGFP	CDB0216K/0251K
Golgi apparatus	Golgi-EGFP	R26R/R26-Golgi-EGFP	CDB0211K/0246K
	Golgi-mCherry	R26R/R26-Golgi-mCherry	CDB0212K/0247K
Membrane	Lyn-Venus	R26R/R26-Lyn-Venus	CDB0219K/0254K
Microtubules	EGFP-Tuba	R26R/R26-EGFP-Tuba	CDB0210K/0245K
	hEMTB-EGFP	R26R/R26-hEMTB-EGFP	CDB0214K/0249K
	EB1-EGFP	R26R/R26-EB1-EGFP	CDB0213K/0248K
Actin cytoskeleton	Venus-Actin	R26R/R26-Venus-Actin	CDB0218K/0253K
	Venus-Moesin	R26R/R26-Venus-Moesin	CDB0217K/0252K
Focal contact	EGFP-Paxillin	R26R/R26-EGFP-Paxillin	CDB0221K/0256K
Nucleus and membrane	H2B-mCherry-2A-EGFP-GPI	R26R/R26-RG	CDB0227K/0237K

List of R26R and R26 reporter lines generated by inserting a series of fluorescent fusion protein cDNAs into the ROSA26 locus. The R26 lines constitutively expressing reporter fusion proteins were established by crossing each R26R line with the EIIa-Cre transgenic line (21)

2.4 Materials for Culture of Postimplantation Embryos in Collagen Gel

1. Medium for collection of postimplantation embryos: DMEM.

2. Sterile rat serum for culture.

3. Reconstruction buffer: 0.183 M HEPES, 0.08 M NaOH.

4. Culture medium: DMEM supplemented with 1 mM β-mercaptoethanol, 1 mM sodium pyruvate, and 100 μM nonessential amino acid.

5. Reconstituted collagen solution: Seven parts Cellmatrix Type I-A collagen gels (Nitta Gelatin Inc., Osaka, Japan), two parts 5× culture medium, and one part reconstruction buffer.

6. Fine tweezers.

3   Methods

3.1 Culturing and Imaging of Preimplantation Embryos

The culturing of preimplantation embryos is already an established technique (2). For the confocal observation, the use of glass-bottom dish is highly recommended because they provide better optical resolution for embryo cultures.

1. Place a drop of KSOM medium in a 35 mm glass-bottom dish. Then overlay the whole dish with mineral oil. Incubate the dish in a CO_2 incubator at 37 °C until use.

2. Collect and transfer embryos to the KSOM drop using a glass pipette with a mouthpiece (Fig. 2a).

3. Observe embryos in 5 % CO_2 at 37 °C in the imaging equipment. Embryos can develop until the blastocyst stage without changing the medium. See note 3.

3.2 Preparation of Collagen Gel for Postimplantation Embryos

1. Collect embryos in DMEM medium at room temperature.

2. Transfer embryos to 50 μl reconstituted collagen solution (see Section 2.4 for ingredients) on a 35 mm glass-bottom culture dish at room temperature.

3. Adjust the orientation of the embryos with tweezers when collagen starts to gel. See note 4. Collagen mixture is maintained as a solution while chilling and as a gel when warmed.

4. Incubate the dish in a 37 °C incubator to complete gelation.

5. Overlay the collagen gels with 150 μl culture medium supplemented with 50 % rat serum.

6. Observe embryos in 5 % CO_2 at 37 °C in the imaging equipment.

4　Notes

1. Selection of mouse lines

   (a) The first important step for live observations of embryonic development is selecting the reporter mouse line to be used in the experiments. This selection process varies among researchers depending on their purpose and experimental design. The researchers can select the color of fluorescent proteins and the subcellular localization of fluorescently labeled fusion proteins. As shown in Table 1, we have established a series of reporter mouse lines for visualizing specific organelles (1).

   Nucleus labeling can be used to trace cell movements and cell lineages, as well as to visualize cell divisions and apoptosis (3). H2B-EGFP and H2B-mCherry were generated for this labeling (4, 5), which mark histone H2B with a green fluorescent protein EGFP or a red fluorescent protein mCherry, respectively. Because these fusion proteins label the chromosomes directly, the daughter cells can easily be identified during cell division, also useful for the tracing of cell lineages (6).

   Mito-EGFP marked mitochondria (7), and Golgi-EGFP and Golgi-mCherry marked both *cis-* and *trans*-Golgi apparatuses (8), respectively. These reporters provide information on the behavior of organelles and the polarity of cells in living embryos.

   When tracking the shape of cells in embryos, there are several useful reporter lines that can be used which include labeling of the cytoskeletal components. Lyn-Venus marks the membrane by myristoylation signals (9). EGFP-Tuba (10) and hEMTB-EGFP (11) mark the microtubules. EB1-EGFP can be used to visualize microtubule plus-ends (12), allowing us to observe the ends of growing microtubules with relatively fast speeds. Venus-Actin (13, 14) and Venus-Moesin (14) mark actin cytoskeleton with the yellowish-green fluorescent protein Venus (15). Focal adhesions of cells can be visualized by EGFP-Paxillin (16).

   (b) Combinations of two colors can be used to visualize two different regions of a cell. For example, a green organelle probe that labels one organelle can be combined with H2B-mCherry that labels chromosomes. This can be done by using embryos obtained from the crossing of two different reporter mouse lines. It is also possible to visualize two kinds of fluorescent markers encoded on a single locus; for example, H2B-mCherry-2A-EGFP-GPI (RG) can simultaneously mark the nucleus and membrane (17).

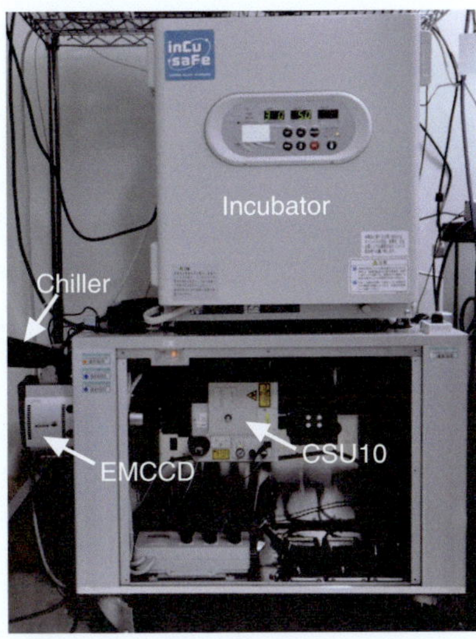

Fig. 1 Image of LCV100 incubation system with CSU. CSU10 spinning disc confocal system is connected to the bottom of the conventional CO_2 incubator to create an optical path. *On the left*, the EMCCD chilled by circulating cold water is used to capture the images

The cDNAs of each reporter are attached to a sequence encoding 2A-peptide, which is then cleaved to make two separate proteins (18). All reporter mouse lines mentioned above are available from RIKEN CDB (http://www.cdb.riken.jp/arg/reporter_mice.html). These reporter fusion fluorescent proteins are expressed under the Rosa 26 locus, which is well known for its ubiquitous expression. For tissue-specific labeling, R26R conditional reporter lines should be crossed with a Cre-transgenic mouse expressing Cre recombinase spatiotemporally in a specific tissue (19, 20).

2. Imaging equipment

    (a) For the live imaging of developing embryos, we need to pay close attention to phototoxicity. Because the fluorescent proteins are only visible by illuminating them with lights of certain wavelengths, we must avoid damaging the cells when illuminating them for observation. Therefore, it is important to carefully consider how to best optimize the imaging equipment.

    (b) We used the Olympus incubation imaging system LCV100 equipped with a CSU10 (YOKOGAWA) and an iXon + EMCCD camera (Andor) (Fig. 1): a $20\times$ objective lens and a triple band filter (YOKOGAWA, transmission: 495–555 nm, 575–635 nm, and 665–775 nm).

Illuminating lasers are 488 and 561 nm. The LCV100 is the combination of a conventional CO_2 incubator and a microscopic system to observe the living cells and embryos cultured in the incubator. This incubation system maintains a more stable environmental condition than a microscope-top incubation chamber to support the development of cultured embryos. Our imaging system is also equipped with a CSU10 spinning disc confocal system to acquire confocal images. An optical Nipkow disc scanner using a rotating disc with pinholes to produce an image with less phototoxicity than the normal line scanning confocal microscopy. The CSU system can illuminate multiple points simultaneously and the energy illuminating each spot is also much weaker. The iXon + EMCCD is an electron multiplying CCD (EMCCD) camera, which can obtain 10–100 times higher sensitivity than a cooled CCD camera. The chiller is equipped to maintain the temperature of the CCD at -90 °C and the sensitivity of the EMCCD camera without increasing background noise. Acquired image data are processed by MetaMorph software (Universal Imaging Corporation). The embryos are cultured in an environment of 5 % CO_2 at 37 °C. Alternatively, a CSU-based microscopy system with a highly stable incubation chamber is also commercially available from Yokogawa (CV1000), in which similar images can be obtained.

3. Imaging of preimplantation embryonic development

   Transcriptional activity of the Rosa26 locus is maternally active, and the maternal storages of RNA and reporter fluorescent proteins accumulate in the oocytes and fertilized eggs. Maternally transmitted reporter(s) can be observed from the oocyte stage. Embryonic genome activation occurs at the 2-cell stage, so the paternal reporter signals are barely detectable at this stage and increase during the 4-cell stage. At the 8-cell stage, most of the paternal reporters observed are the same as the observed maternal reporter genes.

4. Imaging postimplantation embryos

   To observe embryos from stages E5.5 to E7.5, embryos are embedded and cultured in collagen gels (Fig. 2b). Gel-embedded embryos can be used for multipoint observations. Around ten E5.5 embryos can be embedded and cultured in a drop of collagen gel which should be made on a glass-bottom dish.

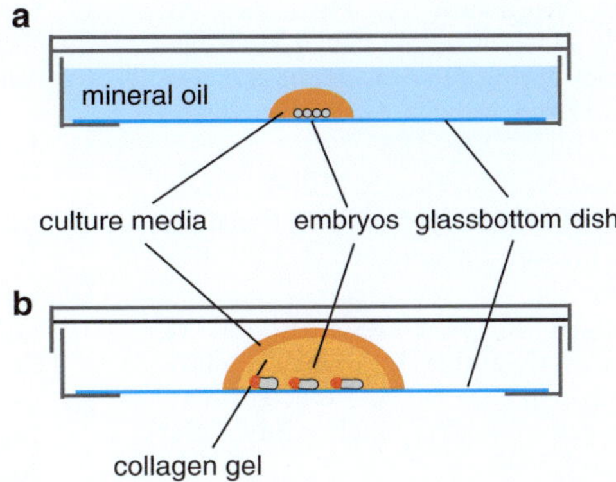

Fig. 2 Schematic drawings of culture dish for live imaging of embryos. (**a**) Preimplantation embryos in a drop of KOSM medium. (**b**) Embedded postimplantation embryos in a collagen gel. In both cases, embryos are settled on glass-bottom dishes

Acknowledgements

The authors acknowledge Dr. Kazuki Nakao, Dr. Hiroshi Kiyonari, Dr. Shioi Go, and Mr. Ken-ichi Inoue for producing reporter mouse lines.

References

1. Abe T, Kiyonari H, Shioi G, Inoue K, Nakao K et al (2011) Establishment of conditional reporter mouse lines at ROSA26 locus for live cell imaging. Genesis 49:579–590

2. Nagy A (2003) Manipulating the mouse embryo: a laboratory manual, 3rd edn. Cold Spring Harbor Laboratory, Cold Spring Harbor, NY, 764 pp

3. Hadjantonakis AK, Papaioannou VE (2004) Dynamic in vivo imaging and cell tracking using a histone fluorescent protein fusion in mice. BMC Biotechnol 4:33

4. Kanda T, Sullivan KF, Wahl GM (1998) Histone-GFP fusion protein enables sensitive analysis of chromosome dynamics in living mammalian cells. Curr Biol 8:377–385

5. Shaner NC, Campbell RE, Steinbach PA, Giepmans BN, Palmer AE et al (2004) Improved monomeric red, orange and yellow fluorescent proteins derived from Discosoma sp. red fluorescent protein. Nat Biotechnol 22:1567–1572

6. Kurotaki Y, Hatta K, Nakao K, Nabeshima Y, Fujimori T (2007) Blastocyst axis is specified independently of early cell lineage but aligns with the ZP shape. Science 316:719–723

7. Rizzuto R, Brini M, Pizzo P, Murgia M, Pozzan T (1995) Chimeric green fluorescent protein as a tool for visualizing subcellular organelles in living cells. Curr Biol 5:635–642

8. Llopis J, McCaffery JM, Miyawaki A, Farquhar MG, Tsien RY (1998) Measurement of cytosolic, mitochondrial, and Golgi pH in single living cells with green fluorescent proteins. Proc Natl Acad Sci U S A 95:6803–6808

9. Teruel MN, Blanpied TA, Shen K, Augustine GJ, Meyer T (1999) A versatile microporation technique for the transfection of cultured CNS neurons. J Neurosci Methods 93:37–48

10. Kimble M, Kuzmiak C, McGovern KN, de Hostos EL (2000) Microtubule organization and the effects of GFP-tubulin expression in dictyostelium discoideum. Cell Motil Cytoskeleton 47:48–62

11. Faire K, Waterman-Storer CM, Gruber D, Masson D, Salmon ED et al (1999) E-MAP-115 (ensconsin) associates dynamically with microtubules in vivo and is not a physiological modulator of microtubule dynamics. J Cell Sci 112(Pt 23):4243–4255

12. Piehl M, Cassimeris L (2003) Organization and dynamics of growing microtubule plus ends during early mitosis. Mol Biol Cell 14:916–925

13. Westphal M, Jungbluth A, Heidecker M, Muhlbauer B, Heizer C et al (1997) Microfilament dynamics during cell movement and chemotaxis monitored using a GFP-actin fusion protein. Curr Biol 7:176–183

14. Iioka H, Ueno N, Kinoshita N (2004) Essential role of MARCKS in cortical actin dynamics during gastrulation movements. J Cell Biol 164:169–174

15. Nagai T, Ibata K, Park ES, Kubota M, Mikoshiba K et al (2002) A variant of yellow fluorescent protein with fast and efficient maturation for cell-biological applications. Nat Biotechnol 20:87–90

16. Petit V, Boyer B, Lentz D, Turner CE, Thiery JP et al (2000) Phosphorylation of tyrosine residues 31 and 118 on paxillin regulates cell migration through an association with CRK in NBT-II cells. J Cell Biol 148:957–970

17. Shioi G, Kiyonari H, Abe T, Nakao K, Fujimori T et al (2011) A mouse reporter line to conditionally mark nuclei and cell membranes for in vivo live-imaging. Genesis 49:570–578

18. de Felipe P, Luke GA, Hughes LE, Gani D, Halpin C et al (2006) E unum pluribus: multiple proteins from a self-processing polyprotein. Trends Biotechnol 24:68–75

19. Soriano P (1999) Generalized lacZ expression with the ROSA26 Cre reporter strain. Nat Genet 21:70–71

20. Srinivas S, Watanabe T, Lin CS, William CM, Tanabe Y et al (2001) Cre reporter strains produced by targeted insertion of EYFP and ECFP into the ROSA26 locus. BMC Dev Biol 1:4

21. Lakso M, Pichel JG, Gorman JR, Sauer B, Okamoto Y et al (1996) Efficient in vivo manipulation of mouse genomic sequences at the zygote stage. Proc Natl Acad Sci USA 93:5860–5865

Methods Molecular Biology (2013) 1052: 109–123
DOI 10.1007/7651_2013_19
© Springer Science+Business Media New York 2013
Published online: 3 May 2013

Live Imaging, Identifying, and Tracking Single Cells in Complex Populations In Vivo and Ex Vivo

Minjung Kang, Panagiotis Xenopoulos, Silvia Muñoz-Descalzo, Xinghua Lou, and Anna-Katerina Hadjantonakis

Abstract

Advances in optical imaging technologies combined with the use of genetically encoded fluorescent proteins have enabled the visualization of stem cells over extensive periods of time in vivo and ex vivo. The generation of genetically encoded fluorescent protein reporters that are fused with subcellularly localized proteins, such as human histone H2B, has made it possible to direct fluorescent protein reporters to specific subcellular structures and identify single cells in complex populations. This facilitates the visualization of cellular behaviors such as division, movement, and apoptosis at a single-cell resolution and, in principle, allows the prospective and retrospective tracking towards determining the lineage of each cell.

Keywords: Embryonic stem cells, Mouse embryo, In vitro culture, *Ex utero* culture, Live imaging, Fluorescent protein, Time lapse, Automated image analysis, Cell tracking

---

1 Introduction

Embryonic stem (ES) cells have been extensively used in order to elucidate the biological mechanisms underlying pluripotency, differentiation, and cellular reprogramming; a deep understanding of these mechanisms could then lay the foundations for an efficient and safe application of these cells for therapeutic purposes. In contrast to the well-known unlimited potential of ES cells, our understanding on their biology is much more limited. Moreover, recent findings regarding heterogeneities in gene expression in stem cell cultures revealed the limitation of static stem cell analysis, which relies only on the status of cells at a certain time point and thus lacks important information about cells' changing their state over time (1, 2). Therefore, extensive analysis of stem cells in vitro

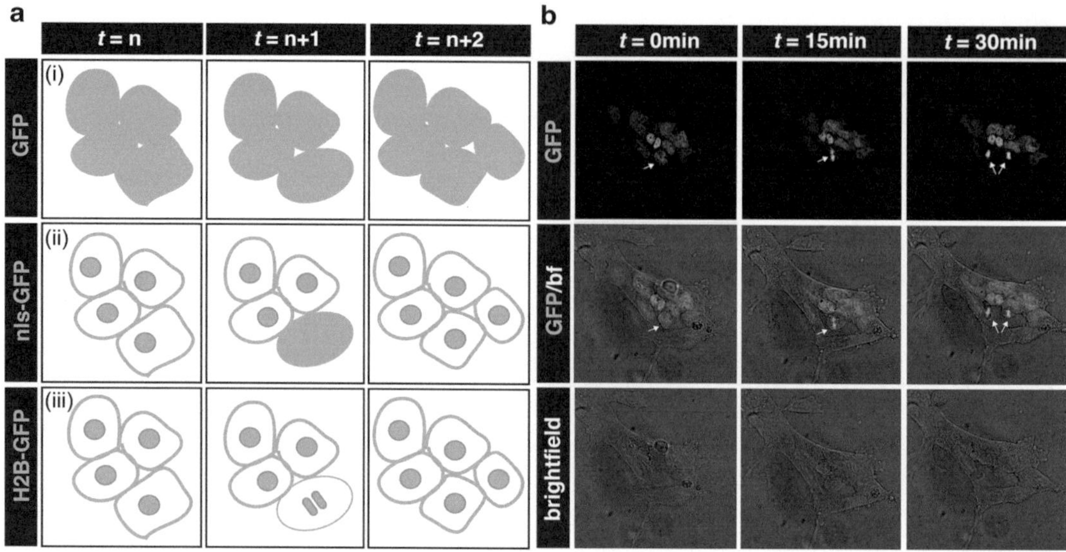

Fig. 1 Fusions of fluorescent proteins (FPs) to human histone H2B label active chromatin and allow for single-cell identification and tracking. (**a**) Schematic cartoons depict cell tracking approaches using FPs. (*i*) Individual cells with cytoplasmic FPs cannot be identified nor tracked. (*ii*) FP fusions to nuclear localization sequences (nls-GFP) can be used to identify and track individual cells by labeling individual nuclei. During cell divisions, however, nuclear envelops break down causing the diffusion of the fluorescent signal to the cytoplasm. Cells cannot be tracked through cell divisions. (*iii*) FP fusions to human histone H2B protein are bound to chromatin structures throughout cell cycle. During cell divisions, the fluorescent signal is condensed as chromatin structures undergo compaction and enables to identify and track the dividing nucleus. (**b**) *Panels* show single time points from a 2D time-lapse movie of a mouse ES cell colony carrying a *Nanog:H2B-GFP* reporter growing in serum + LIF conditions. Images acquired every 15 min. *Arrows* indicate a dividing cell and its daughter cells

(i.e., in live ES cell cultures) as well as in their natural environment in situ (i.e., in developing preimplantation embryos) requires execution in a proper spatial and temporal context for a thorough understanding of stem cell biology.

The remarkable advances in live imaging technologies combined with the use of genetically encoded fluorescent proteins (FPs) have enabled the visualization of stem cells over extensive periods of time in vivo, as well as ex vivo (3–5). The generation of genetically modified FPs that are fused with subcellularly localized proteins such as human histone H2B (H2B) (6), or tagged with localization sequences such as myristoyl anchors (Myr) (7), has made it possible to direct fluorescent protein reporters to specific subcellular structures and observe cellular behaviors such as division, movement, and apoptosis at a single-cell resolution and, in principle, prospectively and retrospectively trace the lineage of each cell (3, 7, 8) (Fig. 1). To this end, the generation of dual-tagged ES cells that express simultaneously fluorescent FPs labeling different subcellular components (e.g., the plasma membrane and the nucleus) has provided high-resolution live imaging of single-cell morphology and easier analysis and tracking of live cells in vitro as well as in vivo (9).

Most importantly, fluorescence microscopy in combination with confocal imaging has allowed the visualization of whole embryos and subcellular structures at a single-cell resolution. Various confocal modalities are available to serve specific purposes in live imaging. Of these, laser point scanning microscopes (e.g., Zeiss LSM 500) are most commonly used and provide higher resolution, with the expense of longer exposure time however, which can cause phototoxicity during embryo imaging. Slit-scanning confocals (e.g., Zeiss LSM5LIVE) or Nipkow-type spinning disc confocals (e.g., Perkin Elmer Ultra View) have faster scanning speed and shorter exposure times, which can compromise imaging resolution. Another option is two-photon laser scanning microscopy (TPLSM), which provides temporal and spatial resolution as well as low phototoxicity for studying preimplantation development (10). In this book chapter, we describe protocols using laser scanning confocal microscope modalities, since they are the most commonly used types.

Long-term imaging, including 4D (3D over time, usually referred to as 3D time-lapse) information, could generate enormous amounts of data that are then further analyzed to address various questions. However, the massive size of such datasets rules out the possibility of employing manual analysis and raises an imperious demand for automated, robust image analysis techniques. The goals of such analysis are (1) to accurately locate cells of interest, (2) to quantitatively characterize their states (e.g., level of reporter expression) and properties, and (3) to robustly, prospectively, and retrospectively track them. To achieve these goals, two major image analysis techniques are applied: cell segmentation (i.e., the process of breaking up a complex multicell image into individual cells) and cell tracking (i.e., where cells are identified in the data from each time point and followed across time points). For a glossary of terms used in the computational analysis of images see (11, 12). Efficient cell segmentation and tracking require the implementation of servers for high-content data storage that allows real-time access to the data as they are being acquired. In this chapter, we describe protocols for imaging live ES cells in vitro, i.e., in stem cell cultures, as well as in vivo, i.e., in the early mouse embryo. We also discuss image segmentation and object tracking techniques that are used to extract spatial and temporal information about the observed cells.

2 Materials

2.1 Medium

2.1.1 Medium and Reagents for Mouse Embryonic Stem Cell Culture

For ES cells, variations in culture medium can induce differential cellular behaviors. Therefore, depending on cell line origin, some culture modifications need to be made in order to prevent their differentiation and maintain their pluripotent state. It is preferable to grow ES cells on gelatinized dishes that contain a layer of adherent feeder cells (mitotically inactivated primary mouse fibroblasts); however, certain ES cell lines (e.g., R1) can be propagated efficiently without feeders, simply on dishes coated with gelatin. We routinely culture ES cells in medium supplemented with serum and the cytokine leukemia inhibitory factor (LIF), which is essential for maintenance of pluripotency. The medium and reagents required for mouse ES cell maintenance are listed below:

1. 75 % DMEM (Invitrogen, 11965-092) + 15 % fetal bovine serum + 1 % Sodium pyruvate (Invitrogen, 11360-070) + 1 % Non-essential amino acid (Invitrogen, 11140-050) + 1 % L-glutamine (Invitrogen, 25030-081) + 1 % Pen/Strep (Invitrogen, 15140-122) + 0.01 % LIF (Millipore, ESG1106).

2. 0.25 % Trypsin–EDTA (Invitrogen, 25300-054).

3. 0.5 % gelatin (Sigma, G9391).

2.1.2 Media and Reagents for Preimplantation Mouse Embryo Culture

Culture and manipulation medium for preimplantation embryos are commercially available. Some very commonly used are the following:

1. M2—for preimplantation embryo manipulation in an atmospheric environment (Millipore, MR-015-D).

2. KSOM—for preimplantation embryo culture in a CO_2 environment (KSOM + AA, Millipore, MR-121-D).

3. Acid Tyrode's solution—for removal of the zona pellucida, the glycoprotein coat encapsulating preimplantation-stage mammalian embryos (Millipore, MR-004-D).

2.2 Culturing Mouse ES Cells and Preimplantation Embryos for Live Imaging

1. Humidified CO_2 incubator.

2. On-stage environmental chamber.

3. Gas mixtures: for ES cells, use 5 % CO_2 in air. For embryos, use 5 % CO_2, 5 % O_2 and 90 % N_2.

4. Plastic cover slips (Fisher, 12-547).

5. 24-well dishes (Becton Dickinson, 353047).

6. 35-mm glass bottom dishes for embryos (MatTek, P35G-1.5-14-C) [NOTE 1].

7. Mouth pipette: Assembled with mouthpiece (HPI Hospital Products Med. Tech., 1501P), latex tubing (Fisherbrand,

22-362-772), and 1,000 µl tip as an adaptor for a Pasteur pipette. A Pasteur pipette is hand-pulled over a flame (Bunsen burner) to a diameter of 1.5× the size of the sample (which in this case are mouse embryos).

8. Glass rods for embryo culture dishes: After a Pasteur pipette is pulled, the broken out thinner portion of the pipette can be used. Forceps are used to cut the thin pulled glass to pieces of ~0.5–1 cm in length (so that they fit in the glass bottom portion of a MatTek dish).

9. Embryo-tested mineral oil (Sigma, M8410).

10. 2 % Bacto Agar + 0.9 % NaCl: The glass bottom of a MatTek dish should be covered with a thin layer of agar to prevent embryos from adhering to the glass.

11. 1 ml Syringe with 26-gauge (Becton Dickinson, 305111 for uterus) or blunt 30-gauge needle (Becton Dickinson, 305106 for oviduct) is used for recovery of embryos by flushing oviducts or uteri. Use a sharpening stone to make a blunt end on the needle.

2.3 Microscopes

1. Stereomicroscope with transmitted light for embryo collection.

2. Laser scanning confocal mounted on an inverted compound microscope with 5×, 10×, 20×, and 40× objectives. 5× or 10× "scanning" objectives are usually used dry and for identification and position samples. 3D time-lapse imaging of preimplantation mouse embryos routinely uses 20× or 40× (with oil) objectives for high-magnification imaging.

3. Computer workstation with image data acquisition and processing software.

3 Methods

3.1 Microscope Setup for Culturing and Imaging Mouse ES Cells and Preimplantation Embryos

The environmental chamber setup on an inverted microscope should recapitulate in utero conditions as closely as possible, by maintaining the correct temperature, humidity, along with the right gas content that is required for mouse ES cell and preimplantation embryo development (Fig. 2).

1. Before preparing ES cells or embryos, turn on the temperature control. Ensure that all the doors of the environmental chamber are tightly closed. It usually takes approximately 15–20′ for the chamber temperature to reach 37.5 °C and stabilize.

2. The CO_2 controller can be turned on just prior to imaging. Calibrate the flow rate before using [NOTE 2].

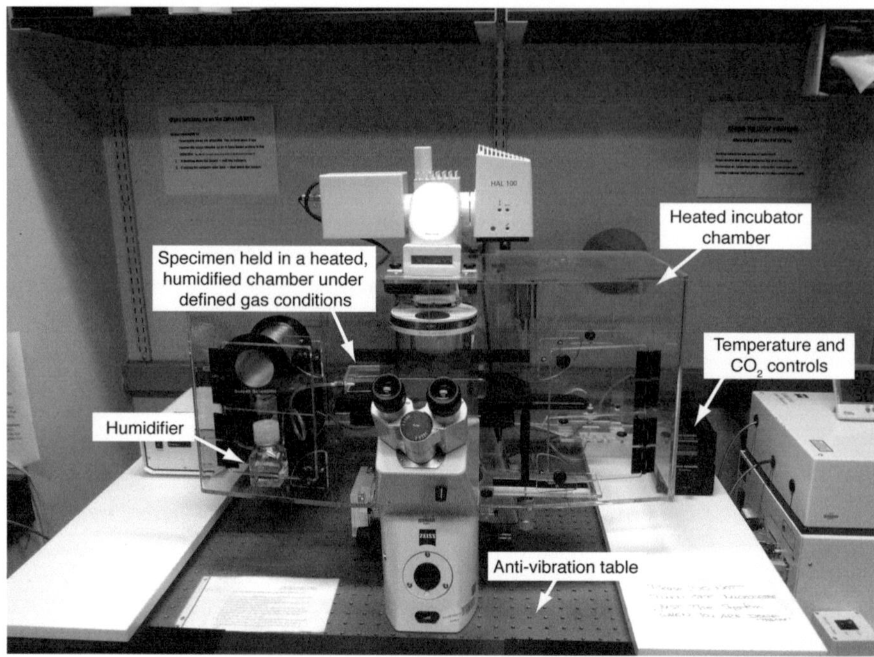

Fig. 2 Microscope setup for live imaging of early mouse embryos as well as in vitro ES cell cultures. Inverted microscopes with environmental chambers provide optimal conditions for *ex utero* and in vitro culture of mouse embryos and mouse ES cells

3.2 Culturing and Imaging ES Cells

3.2.1 Preparation of Culturing Dish for Live Imaging ES Cultures (Fig. 3)

1. Cut a plastic coverslip so that it can fit later into the bottom of a MatTek dish (see shape and dimensions in Fig. 3).

2. Place the cut coverslip in a well of a 24-well dish and coat with gelatin overnight [NOTE 3].

3. Aspirate gelatin and place with 4×10^4 feeders. Once feeders have settled to the bottom of the well (usually after 3–4 h), then seed 4×10^4 ES cells. If feeders are not used, ES cells can be seeded directly onto the gelatinized plastic coverslip.

4. After 12–24 h remove the plastic coverslip with forceps and place it inverted (i.e., with the cells facing downwards) in the bottom of a MatTek Dish. Gently add 3–4 ml of ES cell media so that it fully covers the bottom of the dish [NOTE 4]. Cover the drop completely with mineral oil to prevent evaporation. Place the dish for at least 30 min–1 h in the humidified microscope incubator at 37.5 °C, with 5 % CO_2.

3.2.2 Live Imaging of ES Cell Cultures (Fig. 4)

We try to minimize cell exposure to laser light by reducing laser power and exposure time, decreasing the frequency and/or increasing the size of the optical sections and scan speed. We usually acquire images every 15 min and image stacks of up to 80 μm with 2 μm intervals.

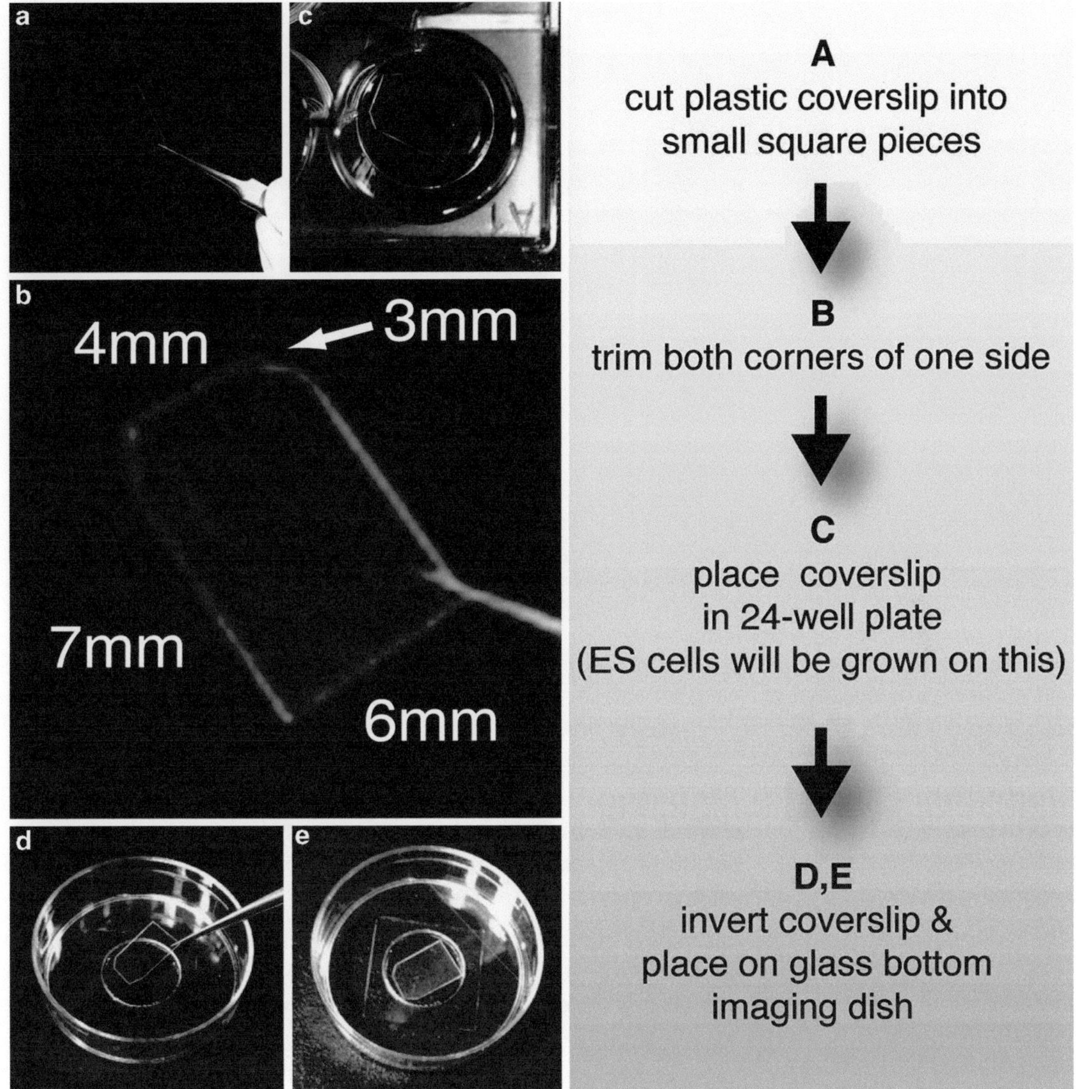

Fig. 3 Preparation of dish for live imaging of ES cell cultures. (**a**) A plastic coverslip is cut to four equal size square pieces. (**b**) Both top corners of each square piece are then cut, resulting into a coverslip of hexagonal shape. (**c**) The coverslip is placed in the bottom of a well of a 24-well dish where ES cells can be cultured on top. (**d**) After ES cells have adhered and grown on top of the coverslip, this is then inverted and (**e**) placed on the bottom of a MatTek dish

3.3 Culturing and Imaging Preimplantation Mouse Embryos

Before implantation, mouse embryos float freely as they have not yet developed any anchor or physical contact in utero. Therefore, in vitro culture of preimplantation embryos should provide closely resembling conditions as those in utero, particularly with respect to the appropriate media, temperature, and gas settings. These conditions are now largely well established, allowing proper development of preimplantation embryos *ex utero*.

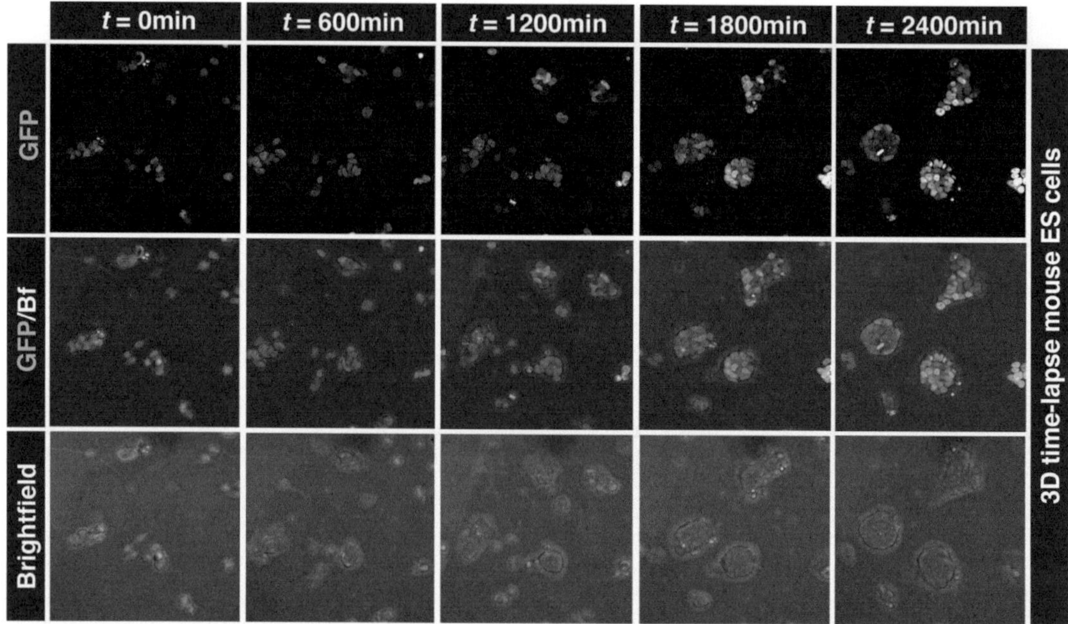

Fig. 4 Time-lapse imaging of ES cell cultures. Our live imaging conditions allow live imaging of individual ES cells as they give rise to stem cell colonies. *Panels* show single time points from a 3D time-lapse movie of an ESC colony carrying a *Nanog:H2B-GFP* reporter growing in serum + LIF conditions

3.3.1 Preparation of Ex Utero Culturing Dish for Mouse Preimplantation Embryos (Fig. 5)

1. The bottom of a MatTek dish for culture should be covered with an extremely thin layer of agar to prevent embryos without zona pellucida from sticking to the glass, as this would limit their proper development [NOTE 5].

2. Place 2–3 glass rods on top of the agar. These serve as holders for the embryos and keep them from floating away from the imaging field, during the time-lapse movie [NOTE 6].

3. Add a drop of KSOM to cover the glass bottom portion of the dish (~400 µl). Cover the drop completely with mineral oil to prevent evaporation. Place the dish in the humidified microscope incubator at 37.5 °C with preimplantation gas mixture (5 % CO_2, 5 % O_2 and 90 % N_2) for at least 15 min so that the medium equilibrates.

3.3.2 Collection of Preimplantation Mouse Embryos

1. Before beginning the isolation of embryos, pre-warm M2 medium and acid Tyrode's solution at 37.5 °C.

2. After sacrificing a pregnant female mouse, dissect out the oviduct or uterus (depending on the desired stage of embryo). Place the tissue in a drop of M2.

3. Flush the oviduct/uterus with pre-warmed M2 medium. Use a 1 ml syringe with a 26-gauge needle (for uteri) or a blunt 30-gauge needle (for oviducts, insert into the infundibulum).

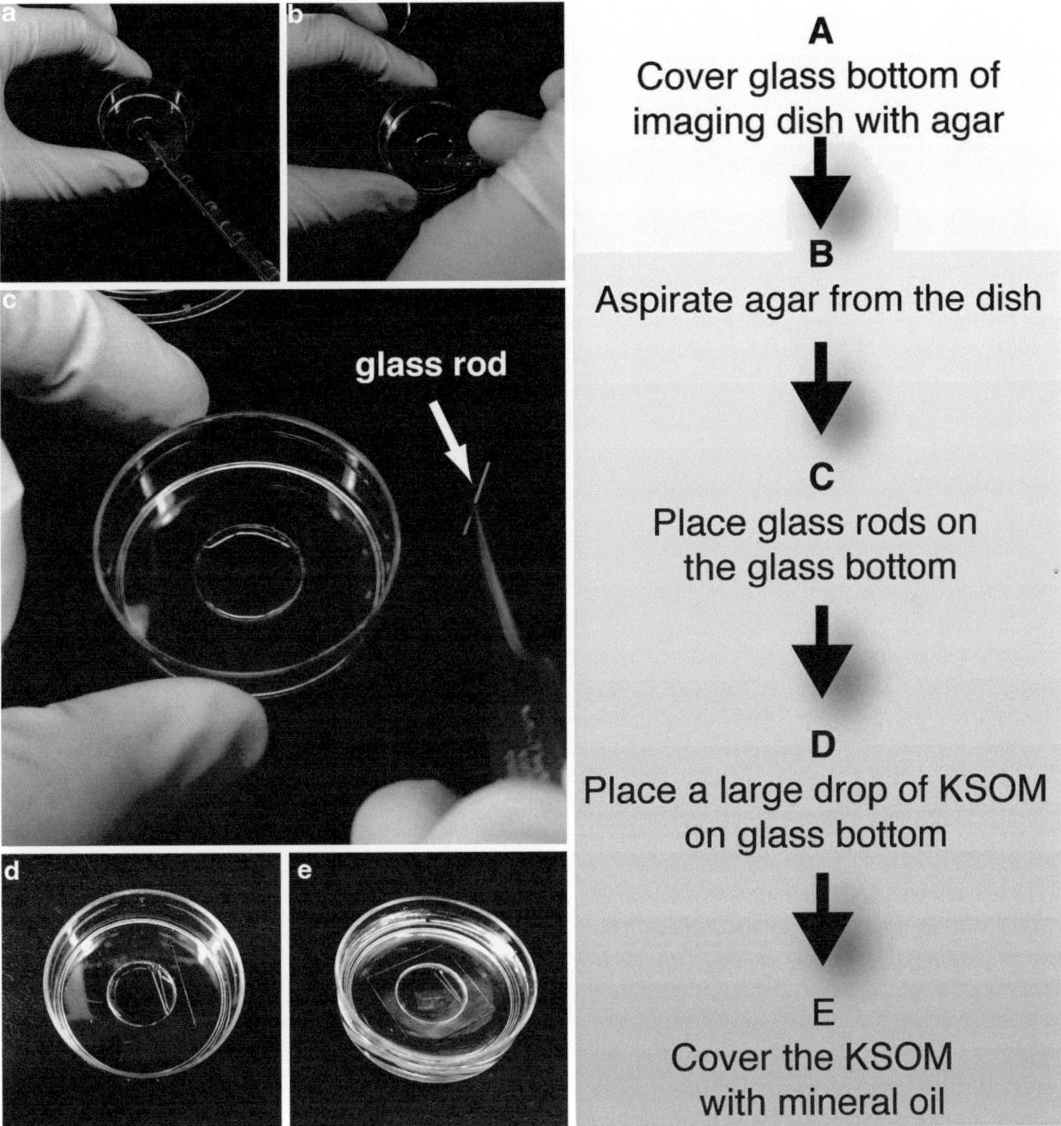

Fig. 5 Preparation of culturing dish for live imaging of preimplantation mouse embryos. (**a**) A glass-bottom dish is covered with agar. (**b**) The dish is tilted and the agar is aspirated off. (**c**) After aspirating off the agar, the dish remains with a thin layer of agar. (**d**) The glass rod is fitted into the glass bottom of the dish. (**e**) A drop of KSOM medium is placed on the bottom of the dish and mineral oil is then added to cover the KSOM drop

4. Collect embryos with a mouth pipette and wash with M2 to remove debris (usually pass through 1–3 drops of M2).

5. Prepare five drops of acid Tyrode's on a 10 cm Petri dish. Place embryos into the first drop briefly, then move to the second drop, and incubate for approximately 2 min. If the zona pellucida has not dissolved after this incubation, embryos should be transferred and incubated successively into the next drops (until the zona pellucida has completely dissolved) [NOTE 7].

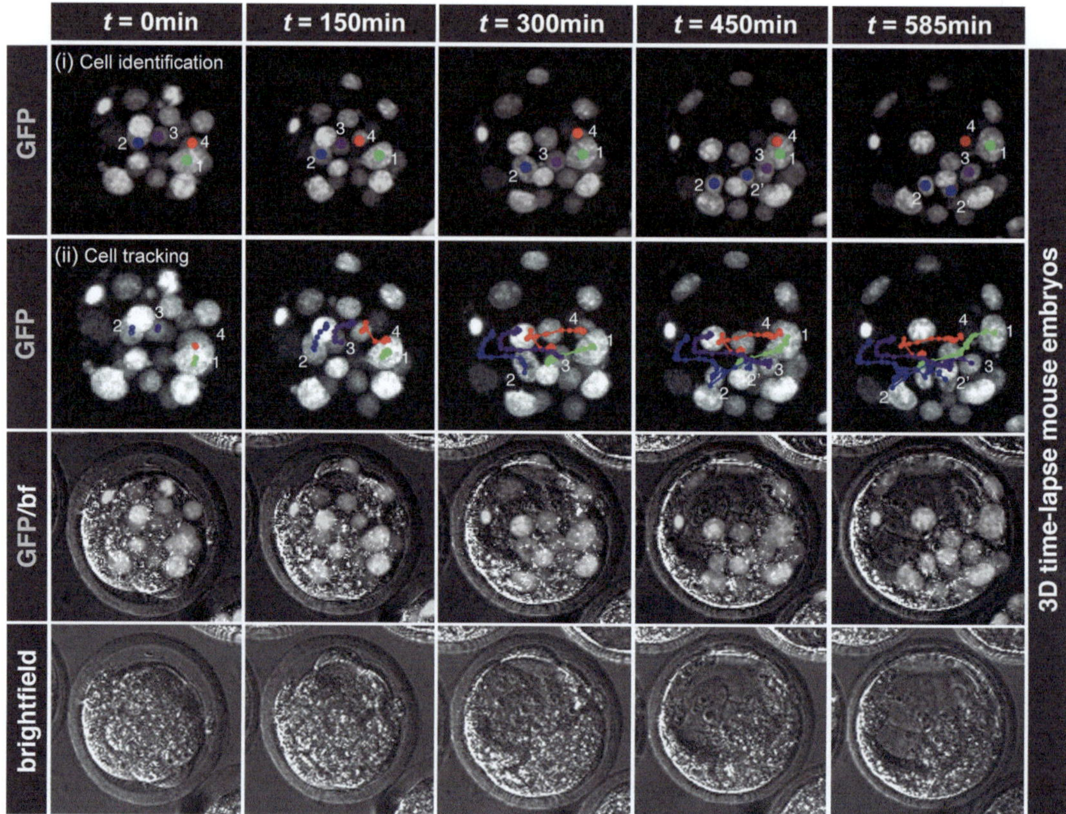

Fig. 6 Time-lapse imaging of preimplantation mouse embryos. Our live imaging conditions allow for imaging of mouse embryos during preimplantation stages. *Panels* show single time points of 3D time-lapse movie of preimplantation embryos carrying a *CAG:H2B-GFP* reporter growing in KSOM medium from morula (~32-cell) to mid-blastocyst (~64-cell) stage. The nuclear-localized GFP signal provides single-cell resolution and facilitates cell tracking. (*i*) Identification of the H2B-GFP signal demarcating individual cells in the living embryo. (*ii*) The movements of individual cells are tracked using the nuclear-localized fluorescent signal. All panels represent 3D reconstructions of *z*-stacks acquired during time-lapse experiments

6. Once the zona pellucida is removed, wash embryos in ~3 consecutive drops of M2.

3.3.3 Live Imaging of Preimplantation Mouse Embryos (Fig. 6)

1. After the zona pellucida has been removed and thoroughly washed, place embryos close to the glass rods and position them on the microscope stage by using the 5× objective. With a mouth pipette, bring the embryos close together so that they can all be imaged in the same field.

2. Set up the parameters for embryo time-lapse imaging. Minimize embryo exposure to laser light by reducing laser power and exposure time, by decreasing the frequency and/or increasing the size of optical section and scan speed [NOTE 8].

Table 1
A selection of useful ImageJ plug-ins for analyzing time-lapse images

Name	Description	Link
Manual tracking	Semi-automated movement and quantification of objects between frames of a temporal stack, in 2D and 3D	http://rsbweb.nih.gov/ij/plugins/track/track.html
MTrackJ	Manual tracking of moving objects in image sequences and measurement of basic track statistics	http://www.imagescience.org/meijering/software/mtrackj/
MTrack2	Manual tracking of moving objects in image sequences	http://valelab.ucsf.edu/~nico/IJplugins/MTrack2.html
Circadian gene expression (CGE)	Quantifying level of reporter expression by tracking individual cells	http://bigwww.epfl.ch/sage/soft/circadian/

3.4 Image Analysis and Tracking

High-quality image segmentation (e.g., the identification of individual cells, by virtue of their nuclear label) is an important prerequisite for performing downstream tasks comprising common image analysis pipelines. However, segmentation can be a daunting task because of cell deformation, irregularity in appearance, debris, imaging artifacts, and, most noticeably, noise and blurring. In practice, the quality of images varies depending on the imaging system, the nature of the sample, and the type of fluorescent reporter used.

Many segmentation methods have been exploited in the context of embryo data analysis. They each differ in their underlying image processing technique: deformable models (e.g., level set) (13, 14), blob or local maximum detection (15–17), watershed (18, 19), and Markov random field (e.g., graph cut) (20). They also differ in channels needed—both the nuclei and the membrane channels are always preferred for analysis, yet quite often the membrane channel is unavailable because of other experimental considerations. When cell shape can be qualitatively characterized, this information could be also incorporated to improve the segmentation quality (15, 21). These methods are mostly adjusted to specific imaging techniques and experimental conditions, often resulting in custom-made image processing methodologies; that, however, obscures a united comparison between these methodologies. Moreover, researchers interested in image analysis of their results implement algorithms in open-source software like ImageJ (http://rsbweb.nih.gov/ij/), Farsight (http://www.farsight-toolkit.org/wiki/Main_Page), Gofigure2 (http://sourceforge.net/), and ilastik (http://ilastik.org/). Since ImageJ was released in the first place, many researchers have been familiar with its usage and keep developing multiple plug-ins that are made available for the community and contain user manuals. The amount of plug-ins for ImageJ keeps on increasing and they are regularly updated (Table 1 shows a list of useful plug-ins for time-lapse imaging analysis). These are relatively simple methods, but most of them are purely manual or semi-automated, which is inapplicable given the huge amount of data generated during time-lapse imaging. Therefore, there is an increasing trend towards developing generic, trainable software frameworks based on machine learning approaches that can interact with biologists to solve a variety of problems (22, 23).

Cell tracking is another challenging problem, mainly due to low temporal resolution, cell deformation, dense population, and, inevitably, segmentation errors (5). Early cell tracking methods were derived from classic tracking methods which exploited industrial video analysis (e.g., surveillance, motion recognition, traffic monitoring), which included level set (24–26), Kalman/particle filter (27, 28), and graph matching methods (29, 30). Realizing their limitations in capturing large displacement, complex mitosis events, and, most importantly, scalability, researchers are now switching to an association-based approach that is usually expressed as (integer) linear programming problem (31, 32). With the help of some powerful commercial solvers such as CPLEX and Gurobi, this approach has proven to be highly efficient in tracking even thousands of cells (21). To avoid tedious parameter tuning, an advanced machine learning technique has been developed, which automatically optimizes the tracking model using biologists' annotations of preferred tracks (33). Some algorithmic software packages are available, but they still need to be adapted and evaluated for application to cell tracking in mouse embryos and lineage reconstructions (30, 31, 33).

In summary, despite many encouraging advances in biomedical image analysis, cell segmentation remains the core problem in most image-processing pipelines. Cell tracking can be performed in a very robust manner by using machine learning and high-dimensional features, provided that high-quality segmentation is available. For this to be achieved, it will require combined efforts from multiple scientific disciplines: higher resolution microscopy from physicists, better reporters from biologists, and more intelligent algorithms from computer scientists should be provided.

4   Notes

1. Confocal images usually require imaging through less than 1.5 mm thick glass-bottom dishes.

2. CO_2 is supplied through a humidifier bottle, which contains ddH_2O. The flow amount and rate of CO_2 are monitored through the bubbles generated in the humidifier. The proper amount of CO_2 supplies should be determined empirically.

For preimplantation embryos, 2–3 bubbles per seconds yield good results.

3. While adding gelatin, the plastic coverslip should not float; in fact, it can be pushed towards the bottom of the well with a tip.

4. ES cell medium must be added very gently so as to ensure that the coverslip is not removed from the bottom of the MatTek dish. If it does get removed, then it can be gently placed back with the help of a tip, taking however care not to disturb the cells.

5. Using a disposable plastic pipette, make a drop of melted agar on the glass bottom portion of the MatTek dish. Immediately afterwards, tilt the dish and aspirate off the agar. At the surrounding edge on the glass bottom, agar can accumulate and form a thicker layer than in the middle. This can be useful to fix the glass rod later.

6. Place the first glass rod across the glass bottom portion. Immobilize the glass rod by pushing one of its ends at the edge of the glass bottom where the agar is thicker. Then, place the second and third rods in close range. After setting up the dish on the microscope stage, place the embryos in between the rods and then reposition the rods in order to stabilize the embryos.

7. When culturing and imaging embryos from the morula stage, removing the zona pellucida is not recommended, since without zona pellucida, embryos tend to aggregate.

8. Optimal imaging conditions (exposure time, laser power, imaging frequency, imaging interval, etc.) should be optimized empirically. For preimplantation embryos, 2 μm imaging slices at 15-min time intervals with a low laser power generally yield good results. We routinely find that 5 % of a 6.1 A power for Argon lasers works well on our systems.

Acknowledgments

We thank Marilena Papaioannou for critical reading and comments on this chapter. Work in our laboratory is supported by the Human Frontiers Science Program (HFSP) and National Institutes of Health (RO1-HD052115 and RO1-DK084391).

References

1. Chambers I et al (2007) Nanog safeguards pluripotency and mediates germline development (Translated from eng). Nature 450 (7173):1230–1234 (in eng)

2. Toyooka Y, Shimosato D, Murakami K, Takahashi K, Niwa H (2008) Identification and characterization of subpopulations in undifferentiated ES cell culture (Translated from eng). Development 135(5):909–918 (in eng)

3. Nowotschin S, Eakin GS, Hadjantonakis AK (2009) Live-imaging fluorescent proteins in mouse embryos: multi-dimensional, multispectral perspectives (Translated from eng). Trends Biotechnol 27(5):266–276 (in eng)

4. Garcia MD, Udan RS, Hadjantonakis AK, Dickinson ME (2011) Live imaging of mouse embryos (Translated from eng). Cold Spring Harb Protoc 2011(4):pdb top104 (in eng)

5. Schroeder T (2011) Long-term single-cell imaging of mammalian stem cells (Translated from eng). Nat Methods 8(4 Suppl):S30–S35 (in eng)

6. Hadjantonakis AK, Papaioannou VE (2004) Dynamic in vivo imaging and cell tracking using a histone fluorescent protein fusion in mice (Translated from eng). BMC Biotechnol 4:33 (in eng)

7. Rhee JM et al (2006) In vivo imaging and differential localization of lipid-modified GFP-variant fusions in embryonic stem cells and mice (Translated from eng). Genesis 44 (4):202–218 (in eng)

8. Hadjantonakis AK, Dickinson ME, Fraser SE, Papaioannou VE (2003) Technicolour transgenics: imaging tools for functional genomics in the mouse (Translated from eng). Nat Rev Genet 4(8):613–625 (in eng)

9. Nowotschin S, Eakin GS, Hadjantonakis AK (2009) Dual transgene strategy for live visualization of chromatin and plasma membrane dynamics in murine embryonic stem cells and embryonic tissues (Translated from eng). Genesis 47(5):330–336 (in eng)

10. McDole K, Xiong Y, Iglesias PA, Zheng Y (2011) Lineage mapping the pre-implantation mouse embryo by two-photon microscopy, new insights into the segregation of cell fates (Translated from eng). Dev Biol 355 (2):239–249 (in eng)

11. Roeder AH, Cunha A, Burl MC, Meyerowitz EM (2012) A computational image analysis glossary for biologists (Translated from eng). Development 139(17):3071–3080 (in eng)

12. Locke JC, Elowitz MB (2009) Using movies to analyse gene circuit dynamics in single cells (Translated from eng). Nat Rev Microbiol 7 (5):383–392 (in eng)

13. Yu W, Lee HK, Hariharan S, Bu W, Ahmed S (2009) Quantitative neurite outgrowth measurement based on image segmentation with topological dependence (Translated from eng). Cytometry A 75(4):289–297 (in eng)

14. Zanella C et al (2010) Cells segmentation from 3-D confocal images of early zebrafish embryogenesis (Translated from eng). IEEE Trans Image Process 19(3):770–781 (in eng)

15. Bao Z et al (2006) Automated cell lineage tracing in Caenorhabditis elegans (Translated from eng). Proc Natl Acad Sci U S A 103 (8):2707–2712 (in eng)

16. Keller PJ, Schmidt AD, Wittbrodt J, Stelzer EH (2008) Reconstruction of zebrafish early embryonic development by scanned light sheet microscopy (Translated from eng). Science 322 (5904):1065–1069 (in eng)

17. Keller PJ et al (2010) Fast, high-contrast imaging of animal development with scanned light sheet-based structured-illumination microscopy (Translated from eng). Nat Methods 7(8):637–642 (in eng)

18. Fernandez R et al (2010) Imaging plant growth in 4D: robust tissue reconstruction and lineaging at cell resolution (Translated from eng). Nat Methods 7(7):547–553 (in eng)

19. Olivier N et al (2010) Cell lineage reconstruction of early zebrafish embryos using label-free nonlinear microscopy (Translated from eng). Science 329(5994):967–971 (in eng)

20. Lou X, Kothe U, Wittbrodt J, Hamprecht FA (2012) Learning to segment dense cell nuclei with shape prior. In: IEEE conference on computer vision and pattern recognition. Providence, RI, US

21. Lou X et al (2011) Digital embryo lineage tree reconstructor. In: IEEE international sysmposium on biomedical imaging: from Nano to Macro. Chicago, IL, US

22. Carpenter AE et al (2006) Cell profiler: image analysis software for identifying and quantifying cell phenotypes (Translated from eng). Genome Biol 7(10):R100 (in eng)

23. Sommer C, Strahle C, Kothe U, Hamprecht FA (2011) Interactive learning and segmentation toolkit. In IEEE international symposium on biomedical imaging: from Nano to Macro. Chicago, IL, US

24. Dufour A et al (2005) Segmenting and tracking fluorescent cells in dynamic 3-D microscopy with coupled active surfaces (Translated from eng). IEEE Trans Image Process 14 (9):1396–1410 (in eng)

25. Li K et al (2008) Cell population tracking and lineage construction with spatiotemporal context (Translated from eng). Med Image Anal 12(5):546–566 (in eng)

26. Dzyubachyk O, van Cappellen WA, Essers J, Niessen WJ, Meijering E (2010) Advanced level-set-based cell tracking in time-lapse fluorescence microscopy (Translated from eng). IEEE Trans Med Imaging 29(3):852–867 (in eng)

27. Meijering E, Dzyubachyk O, Smal I, van Cappellen WA (2009) Tracking in cell and developmental biology (Translated from eng). Semin Cell Dev Biol 20(8):894–902 (in eng)

28. Smal I, Draegestein K, Galjart N, Niessen W, Meijering E (2008) Particle filtering for multiple object tracking in dynamic fluorescence microscopy images: application to microtubule growth analysis (Translated from eng). IEEE Trans Med Imaging 27(6):789–804 (in eng)

29. Li F, Zhou X, Ma J, Wong ST (2010) Multiple nuclei tracking using integer programming for quantitative cancer cell cycle analysis (Translated from eng). IEEE Trans Med Imaging 29 (1):96–105 (in eng)

30. Liu M et al (2011) Adaptive cell segmentation and tracking for volumetric confocal microscopy images of a developing plant meristem (Translated from eng). Mol Plant 4 (5):922–931 (in eng)

31. Padfield D, Rittscher J, Roysam B (2011) Coupled minimum-cost flow cell tracking for high-throughput quantitative analysis (Translated from eng). Med Image Anal 15 (4):650–668 (in eng). Chicago, IL, US

32. Kanade T et al (2011) Cell image analysis: algorithms, systems and applications. In: IEEE workshop on application of computer visions (WACV).

33. Lou X, Hamprecht FA (2011) Structured learning for cell tracking. In: Neural information processing systems (NIPS). Granada, Spain

Methods Molecular Biology (2013) 1052: 125–141
DOI 10.1007/7651_2013_17
© Springer Science+Business Media New York 2013
Published online: 4 June 2013

Quantitative Evaluation of Stem Cell Grafting in the Central Nervous System of Mice by In Vivo Bioluminescence Imaging and Postmortem Multicolor Histological Analysis

Kristien Reekmans, Nathalie De Vocht, Jelle Praet, Debbie Le Blon, Chloé Hoornaert, Jasmijn Daans, Annemie Van der Linden, Zwi Berneman, and Peter Ponsaerts

Abstract

Stem cell transplantation in the central nervous system (CNS) is currently under intensive investigation as a novel therapeutic approach for a variety of brain disorders and/or injuries. However, one of the main hurdles at the moment is the lack of standardized procedures to evaluate cell graft survival and behavior following transplantation into CNS tissue, thereby leading to the publication of confusing and/or conflicting research results. In this chapter, we therefore provide validated in vivo bioluminescence and postmortem histological procedures to quantitatively determine: (a) the survival of grafted stem cells, and (b) the microglial and astroglial cell responses following cell grafting.

Keywords: Neural stem cells, Mesenchymal stromal cells, eGFP, Luciferase, Cell transplantation, Brain, Bioluminescence imaging, Histology, Microglia, Astrocytes

1   Introduction

Stem cell therapy is widely expected to become a treatment strategy for a variety of disorders, with many preclinical animal studies demonstrating the potential beneficial effect of stem cell grafting in disease models. However, although several potential working mechanisms have been proposed, such as immunomodulation, stimulation of endogenous regeneration-inducing mechanisms, and/or direct cell replacement, none of the proposed mechanisms has been demonstrated to be the exclusive mode of action.

With regard to stem cell transplantation as a treatment for neurological disorders, two types of stem cells, neural stem cells

(NSC) and bone marrow-derived stromal cells (BMSC), have been intensely investigated over the past years (1–5). NSC are multipotent progenitor cells present in both the developing and in the adult central nervous system (CNS). Although they have a strict spatiotemporal occurrence in vivo, they can be easily isolated and cultured from various sources ex vivo. Moreover, their ability to differentiate towards neurons, astrocytes, and oligodendrocytes (observed in both ex vivo neurosphere-cultured NSC and monolayer-cultured NSC populations) has made them interesting candidates to become part of novel treatment strategies for CNS disorders. On the other hand, BMSC have been ascribed immune-modulating properties, which makes them interesting candidates for neuroinflammatory disorders. Moreover, BMSC have been extensively investigated as cellular tools to deliver therapeutic factors to the CNS (6). However, although both NSC and MSC have proven to be successful in improving clinical outcome in models of neuro-inflammation and -degeneration following intravenous, intraventricular, and intra-tissue grafting, the exact contribution of these stem cells remains elusive as to date only few studies are able to provide a link between cell fate and clinical outcome. Moreover, only a limited amount of research is available that describes the impact of stem cell transplantation in the CNS on the host endogenous environment (7–10). This is striking as the host environment is a major player in determining the outcome and the success of stem cell-based therapeutic strategies (11).

In this chapter, we describe standardized procedures for isolation, culture, and genetic modification of NSC and BMSC (4, 5, 9, 10, 12, 13), in order to: (a) longitudinally assess stem cell graft survival using in vivo bioluminescence imaging, (b) to determine cell graft survival by histological analysis, and (c) to determine endogenous microglial and astroglial brain immune responses, following implantation in CNS of immune-competent mice. This will allow a better understanding of the events following cell implantation and will further contribute to determining the exact mode of action. Moreover, this will further enhance therapeutic options for CNS disorders as direct cell implantation into the CNS with the aim to produce therapeutic factors or direct cell replacement is currently hindered by strong endogenous immune response (9, 10).

2 Materials

2.1 Animals

1. Wild-type FVB/NCrl inbred mice—Charles River (strain code 207)

2.2 Products

1. Phosphate buffered saline (PBS)—10× PBS liquid (Gibco, cat nr: 14080048) diluted to 1× in demineralized water.

2. Accutase—Sigma Aldrich (cat nr: A6964).

3. Bovine fibronectin—R&D Systems (cat nr: 1030-FN). For fibronectin coating: add 5 μg/ml of fibronectin diluted in PBS overnight (or longer) at 4 °C.

4. Trypsin–EDTA—Life Technologies (cat nr: 25300054).

2.3 Cell Isolation: Products

1. Anesthesia: mixture of isoflurane—4 % (Isoflo®, cat nr: 05260-05), oxygen (0.5 l/min), and nitrogen (1 l/min).

2. Isolation medium: PBS supplemented with 100 U/ml penicillin (Life Technologies, cat nr: 15140148) and 100 mg/ml streptomycin (Life Technologies, cat nr: 15140148).

3. Dissociation medium: 0.2 % collagenase A (Roche, cat nr: 10103578001)/DNase-I (2000 Kunitz units/50 ml—Sigma Aldrich, cat nr: D4263) solution in PBS.

2.4 Cell Culture: Medium

1. Neural expansion medium (NEM) consists of Neurobasal A medium (Life Technologies, cat nr: 10888022), supplemented with 10 ng/ml epidermal growth factor (EGF, ImmunoTools, cat nr: 11343406), 10 ng/ml human fibroblast growth factor-2 (hFGF-2, ImmunoTools, cat nr: 11343623), 100 U/ml penicillin (Life Technologies, cat nr: 15140148), 100 mg/ml streptomycin (Life Technologies, cat nr: 15140148), 0.5 μg/ml amphotericin B (Life Technologies, cat nr: 15290026), and 1 % modified N2 supplement. The modified N2 supplement consists of DMEM/F12 medium (Gibco, cat nr: 11320074) supplemented with 7.5 mg/ml bovine serum albumin (BSA, Life Technologies, cat nr: 15260037), 2.5 mg/ml insulin (Sigma Aldrich, cat nr: I1882), 2 mg/ml apo-transferrin (Sigma Aldrich, cat nr: T1147), 0.518 μg/ml sodium selenite (Sigma Aldrich, cat nr: S5261), 1.6 mg/ml putrescine (Sigma Aldrich, cat nr: P5780), and 2 μg/ml progesterone (Sigma Aldrich, cat nr: P8783).

2. Complete isolation medium (CIM) consists of RPMI-1640 medium (Life Technologies, cat nr: 61870-010) supplemented with 8 % horse serum (HS, Life Technologies, cat nr: 1605-122), 8 % fetal calf serum (FCS, Hyclone, cat nr: 10270-106), 100 U/ml penicillin (Life Technologies, cat nr: 15140148), 100 mg/ml streptomycin (Life Technologies, cat nr: 15140148), and 1.25 mg/ml amphotericin B (Life Technologies, cat nr: 15290026).

3. Complete expansion medium (CEM) consists of Iscove modified Dulbecco's medium (IMDM, Cambrex, cat nr: BE12-722 F) supplemented with 8 % horse serum (HS, Life Technologies, cat nr: 1605-122), 8 % fetal calf serum

Table 1
Overview of antibodies used for flow cytometric analysis

Primary AB	Conjugation	Company	Secondary AB	Company
α-Sca-1	PE	eBioscience (12/5981-82)		
α-CD45	PE	BD (553081)		
α-MHCII	PE	eBioscience (12/5321-82)		
α-CD117	FITC	eBioscience (11/1171-82)		
α-MHCI	FITC	BD (553570)		
α-CD31	FITC	eBioscience (11/0311-82)		
α-CD106	FITC	eBioscience (11/1061-82)		
α-A2B5		Millipore (MAB312R)	Goat α-mouse (FITC) Rat α-mouse(PE)	AbD Serotec (Star86F) Jackson Immuno (115-116-075)
α-NCAM		Millipore (MAB310)	Goat α-rat (PE)	Jackson Immuno (112-116-143)

(FCS, Hyclone, cat nr: 10270-106), 100 U/ml penicillin (Life Technologies, cat nr: 15140148), 100 mg/ml streptomycin (Life Technologies, cat nr: 15140148), and 1.25 mg/ml amphotericin B (Life Technologies, cat nr: 15290026).

2.5 Antibodies

1. For flow cytometric characterization of NSC and BMSC. All primary antibodies (AB), as overviewed in Table 1, are reactive to mouse antigens unless stated otherwise and are diluted to a final concentration of 1 μg/100 ml.

2. For histological analysis of transplanted NSC and BMSC. All primary antibodies (AB) are anti-mouse unless stated otherwise. Table 2 provides an overview of validated antibodies and their recommended dilution.

3. Nuclear staining was performed using TOPRO3 deep red stain (diluted 1/200 in TBS)—Life Technologies (cat nr: T3605).

2.6 Flow Cytometric Analysis: Products

1. Gelred (1× final concentration)—Biotum
2. Epics XL-MCL analytical flow cytometer—Beckman Coulter
3. FlowJo Software version 7.2.2—FlowJo

2.7 In Vitro Luminescence Assay: Products

1. White flat bottom plates—Corning Life Sciences (cat nr: 3912)
2. Bright-Glo Luciferase assay system—Promega (cat nr: E2610)
3. Envision (Wallac) 2103 Mutilabel Reader—Perkin Elmer

Table 2
Overview of antibodies used for immunofluorescent histological analysis

Primary AB	Dilution	Company	Secondary AB	Dilution	Company
α-Iba1	1/200	Wako (019-19714)	Donkey α-rabbit AF555	1/500	Life Technologies (A31572)
α-GFAP	1/400	Millipore (MAB360)	Goat α-mouse AF555	1/1,000	Life Technologies (A21127)
α-S100B	1/200	Abcam (ab52642)	Donkey α-rabbit AF555	1/500	Life Technologies (A31572)

2.8 Cell Transplantation: Products

1. Anesthesia: ketamin (80 mg/kg—Pfizer) + xylazin (16 mg/kg—Bayer Health care) mixture
2. Micro-injection pump—Kd Scientific (cat nr: KDS100)
3. Syringe—Hamilton (cat nr: 7635-01)
4. 30-gauge needle—Hamilton (cat nr: 7762-03)
5. 0.9 % NaCl solution—Baxter

2.9 Bioluminescence Imaging: Products

1. Anesthesia: mixture of isoflurane—3 % induction, 1.5 % maintenance (Isoflo®, cat nr: 05260-05), and oxygen (0.5 l/min)
2. In vivo real-time φ-imager system—Biospace
3. D-luciferin (150 mg/kg body weight dissolved in PBS)—Promega Benelux (cat nr: E1601)
4. Analysis software: M3 Vision—Biospace

2.10 Quantitative Histological Analysis: Products

1. Anesthesia: mixture of isoflurane—4 % (Isoflo®, cat nr: 05260-05), oxygen (0.5 l/min) and nitrogen (1 l/min)
2. Fixation: 4 % paraformaldehyde dissolved in PBS—Merck (cat nr: 1.04005.1000)
3. Sucrose gradients: 5, 10, and 20 mg of sucrose (Applichem, cat nr: A1125) dissolved in 100 ml of demineralized water to obtain a 5, 10, and 20 % sucrose solution
4. Micron HM5000 cryostat—Prosan (cat nr: HM525)
5. BX51 fluorescence microscope equipped with an Olympus DP71 digital camera—Olympus
6. Image analysis software:

   NIH ImageJ analysis software—Image J

   Adobe Photoshop C56

   Immunofluorescence analysis software—TissueQuest (TissueGnostics GmbH)
7. Dako pen—Dako (cat nr: S200230)

8. TBS: 15.3 g of NaCl (VWR, cat nr: 1.06404.0500) dissolved in 1.8 l demineralized water supplemented with 200 ml TRIS solution. Thereafter, add four drops of 12 N HCl (VWR, cat nr: 1.00319)

9. TRIS solution: 6.2 g of TRIS (VWR, cat nr: 1.08382.1000) dissolved in 50 ml of demineralized water supplemented with 37 ml of 1 N HCl (VWR, cat nr: 1.00319)

10. Permeabilization buffer: 0.1 % Triton-X dissolved in TBS—Sigma Aldrich (cat nr: X100)

11. Blocking serum: composed of 20 % serum of the host animal of the secondary antibody in TBS

    Goat serum (Jackson Immunoresearch, cat nr: 005-000-121) for GFAP staining

    Donkey serum (Jackson Immunoresearch, cat nr: 017-000-121) for Iba1 and S100B staining

12. Antibody dilution buffer: 10 % milk powder dissolved in TBS

13. Nuclear stain: TOPRO-3 deep red stain diluted 1/200 in TBS—Life Technologies (cat nr: T3605)

---

3 Methods

3.1 Culture and Characterization of NSC and BMSC of Mouse Origin

3.1.1 Neural Stem Cell Culture

1. Anaesthetize a pregnant female WT FVB mouse (E12-E16 embryos) deeply by inhalation of an isofluorane/oxygen/nitrogen mixture for 2 min.

2. Sacrifice the mouse by cervical dislocation.

3. Desinfect the abdomen with 70 % Ethanol and pull back the skin to expose the peritoneum.

4. Cut open the peritoneal wall to expose the uterine horns.

5. Remove the uterine horns and place them in a petridish containing ice-cold isolation medium.

6. Cut open each embryonic sac and place the embryos into a new petridish containing ice-cold isolation medium.

7. Dissect the embryonic forebrain from each embryo and place them in separate 15 ml tubes containing 5 ml of ice-cold isolation medium.

8. Place each embryonic forebrain in a clean petridish containing 5 ml of ice-cold isolation medium.

9. Mince the tissues with two scissors for 5 min.

10. Place the minced tissues in 15 ml tubes and centrifuge for 5 min at $129 \times g$.

11. Add 2 ml of dissociation medium to each tube and incubate for 1.5 h at 37 °C in a shaking water bath.

12. After incubation add 8 ml of NEM and centrifuge for 5 min at $129 \times g$.

13. Resuspend the pellets in 10 ml of NEM.

14. Plate out the obtained cell populations in T25 culture flasks.

15. Every 2–3 days add 10 ng/ml of the growth factors EGF and FGF-2 until neurospheres are formed.

16. Remove medium from the flasks and place in a 15 ml tube.

17. Centrifuge for 5 min at $129 \times g$.

18. Add 5 ml of accutase to the pellet and incubate for 5 min at 37 °C.

19. Add 5 ml of NEM medium to the tube and centrifuge for 5 min at $129 \times g$.

20. Add 10 ml of NEM medium to the pellet and plate out the cells in fibronectin-coated (5 µg/ml in PBS) T25 culture flasks.

21. Following 24 h of culture, remove nonadherent cells by replacing medium with 10 ml fresh NEM medium.

22. Every 2–3 days add 10 ng/ml of the growth factors EGF and FGF-2 until 90 % confluency is reached.

For routine cell cultures, NEM is replaced every 3–4 days and NSC cultures are split 1/5 every 7 days according to the following procedure:

23. Remove medium from the T25 culture flasks.

24. Add 3 ml of accutase to the culture flasks and incubate for 5 min at 37 °C.

25. Add 7 ml of NEM to the culture flasks and place the medium in a 15 ml tube.

26. Centrifuge for 5 min at $129 \times g$.

27. Resuspend the pellet in 1 ml of fresh NEM.

28. Replate the cells in a new fibronectin-coated T25 flask in 10 ml NEM.

3.1.2 Bone Marrow-Derived Stromal Cell Culture

1. Anaesthetize a 3-week-old male WT FVB mouse deeply by inhalation of an isoflurane/oxygen/nitrogen mixture for 2 min.

2. Sacrifice the mouse by cervical dislocation.

3. Remove the tibias and femurs of the mouse.

4. Flush the tibias and femurs with ice-cold isolation medium and collect the bone marrow in 15 ml tubes (one tube per mouse) containing ice-cold isolation medium.

5. Wash the harvested bone marrow twice with ice-cold PBS.

6. Plate out the obtained cell populations in T75 flasks (one flask per mouse) in 20 ml of CIM.

7. Following 24 h of culture, remove nonadherent cells by replacing medium with 20 ml of fresh CIM.

8. Replace medium with fresh CIM every 3–4 days until 90 % of confluency is reached (takes approximately 2 weeks).

For routine cell culture, BMSC cultures were split 1/3 every 5–7 days according to the following procedure.

9. Remove medium from the T75 culture flasks.

10. Add 5 ml of trypsin–EDTA to the culture flasks and incubate for 5 min at 37 °C.

11. Add 10 ml of CEM (CIM for passage 1) to the culture flasks and place the medium in a 15 ml tube.

12. Centrifuge for 5 min at 453 × g.

13. Resuspend the pellet in 1 ml of fresh CEM (CIM for passage 1).

14. Replate the cells in a new T75 flask in 20 ml CEM (CIM for passage 1).

3.1.3 Flowcytometric Characterization of NSC and BMSC

Before antibody staining, harvest NSC and BMSC according to the procedures used for routine cell culture (use PBS instead of culture medium). Thereafter wash the cells twice with PBS and resuspend the cells in PBS at a concentration of 1×10^6 cells/ml.

Antibody Staining Using Directly Labeled Primary Antibodies

1. Add 100 μl of cell suspension to a 5 ml tube

2. Add 1 μg of each primary antibody to 5 ml tubes containing cell suspension

3. Incubate for 30 min at 4 °C

4. Wash the cells twice with PBS

5. Resuspend in 1 ml PBS and analyze directly by flow cytometry

Antibody Staining Using Unlabeled Primary Antibodies

1. Add 100 μl of cell suspension to a 5 ml tube

2. Add 1 μg of each primary antibody to 5 ml tubes containing cell suspension

3. Incubate for 30 min at 4 °C

4. Wash the cells twice with PBS and resuspend in 100 μl PBS containing 1 μg of secondary antibody

5. Incubate for 20 min at 4 °C

6. Wash the cells twice with PBS

7. Resuspend in 1 ml PBS and analyze directly by flow cytometry

For all measurements, cell viability was assessed through addition of 1 μl GelRed to the cell suspension immediately before flow cytometric analysis. At least 10,000 cells were analyzed per sample and flow cytometry data were analyzed using FlowJo software.

3.1.4 Expected Outcome

Figure 1 provides representative results regarding culture and characterization of NSC and BMSC from WT FVB mice.

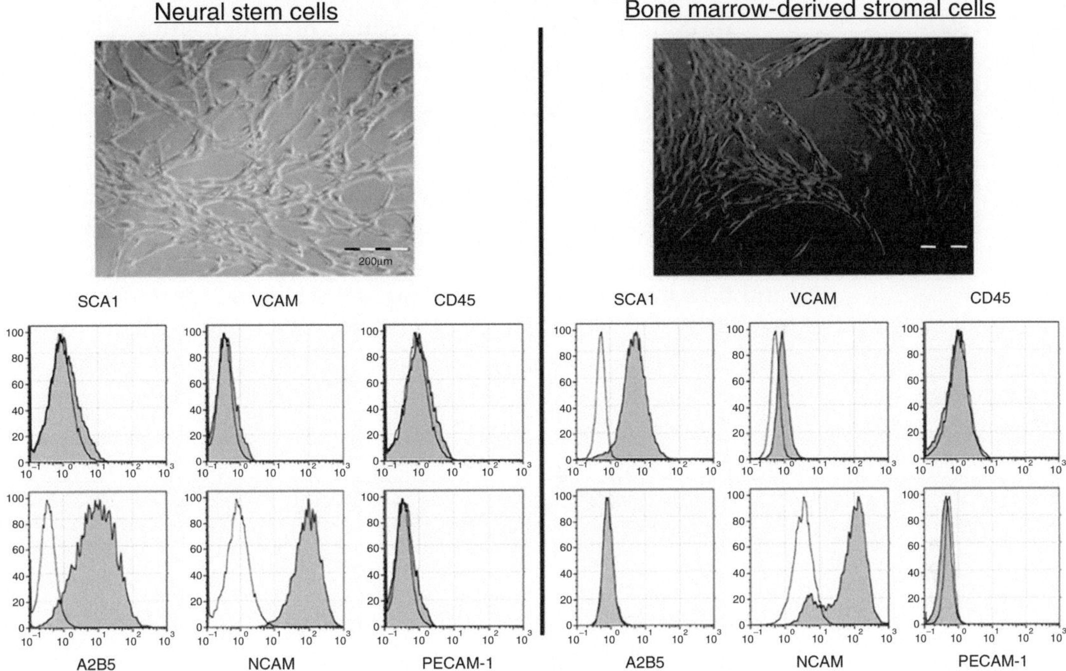

Fig. 1 Cell culture and characterization of neural stem cells and bone marrow-derived stromal cells from WT FVB mice. *Upper figures*: Representative pictures of cultured NSC and BMSC under phase contrast microscopy. *Left*: adherent culture of NSC. *Right*: adherent culture of BMSC. Scale bars: 200 µm. *Lower histogram overlays*: Representative data of flow cytometric analysis of expression patterns of membrane proteins on NSC (*left*) and BMSC (*right*). *Open histogram*: control. *Filled histogram*: specific antibody staining for SCA1, VCAM, CD45, A2B5, NCAM, and PECAM-1. NSC demonstrate expression of membrane proteins A2B5 and NCAM, but no expression of SCA1, VCAM, CD45, and PECAM-1, whereas BMSC demonstrate expression of membrane proteins SCA1, VCAM, and NCAM, but no expression of CD45, A2B5, and PECAM-1

3.2 Introducing eGFP and Luciferase Reporter Gene Expression in Stem Cells via Lentiviral Transduction

3.2.1 Lentiviral Vector Transduction of NSC and BMSC

1. Harvest NSC and BMSC according to the procedures used for routine cell culture.

2. Seed NSC and BMSC in a 24-well plate at a concentration of 25,000 cells per well (NSC are seeded in fibronectin-coated 24-wells) in 1 ml NEM/CEM.

3. Following 24 h of culture, remove medium and add 300 µl fresh NEM/CEM.

4. Add lentiviral vector expressing the eGFP–IRES–Luc cassette at the recommended concentration.

5. Incubate the cells for 48 h at 37 °C.

6. Wash the cells twice with NEM/CEM and replace medium for a period of 3–5 days.

Cells were subcultured at least four times in NEM/CEM according to the procedures used for routine cell culture, whereafter transduction efficiency was determined by flow cytometry.

3.2.2 Flow Cytometric Analysis of eGFP Transgene Expression in NSC and BMSC

1. Harvest eGFP–Luc lentiviral vector-transduced NSC and BMSC according to the procedures used for routine cell culture.

2. Wash the cells once with PBS.

3. Resuspend the cells in 1 ml PBS and analyze directly by flow cytometry.

For all measurements, cell viability was assessed through addition of 1 µl GelRed to the cell suspension immediately before flow cytometric analysis. At least 5,000 cells were analyzed per sample and flow cytometry data were analyzed using FlowJo software.

3.2.3 Obtaining Clonal eGFP–Luciferase-Expressing NSC and BMSC Lines via Limiting Dilution

1. Harvest eGFP–Luc lentiviral vector-transduced NSC and BMSC according to the procedures used for routine cell culture.

2. Resuspend NSC and BMSC at a concentration of 1 cell/75 µl of NEM/CEM.

3. Add 75 µl to each well of a 96-well plate (fibronectin-coated for NSC).

4. Following 3 days of culture add 50 µl of fresh NEM/CEM to each well.

5. After 6–7 days of culture, remove culture medium and replace with 100 µl of fresh NEM/CEM.

6. Split the cells in a 24 well when 70 % confluency is reached according to the procedures used for routine cell culture.

Thereafter, subculture the cells in NEM/CEM according to the procedures used for routine cell culture in 6 wells and then T25/T75 culture flasks. Check the clonicity and eGFP expression of the clonal NSC and BMSC cell cultures with flow cytometry according to Section 3.2.2.

7. Harvest clonal eGFP–Luc lentiviral vector-transduced NSC and BMSC according to the procedures used for routine cell culture.

8. Resuspend NSC and BMSC at a concentration of 100,000 cells per 100 µl.

9. Add 100 µl of cell suspension into the well of a white 96-well plate.

10. Add 100 µl of Bright-Glo Luciferase substrate to the cell suspension, according to the manufacturer's instructions.

11. Analyze using an Envision 2103 Multilabel reader or a real-time φ-imager system and express the luminescence signal as relative light units (RLU) or photons/s/steradian/cm^2.

Include an eGFP–Luciferase negative control NSC and BMSC line to determine the background luminescence signal during each analysis.

3.2.4 Expected Outcome

Figure 2 provides representative results regarding lentiviral vector transduction of NSC and BMSC with eGFP and Luciferase reporter genes.

3.3 Intracerebral Transplantation of Stem Cells in the Mouse Brain

3.3.1 Cell Preparation for Transplantation Experiments

1. Harvest eGFP–Luc lentiviral vector-transduced NSC and BMSC according to the procedures used for routine cell culture

2. Wash the cells twice with PBS

3. Resuspend the cells in PBS at a final concentration of 10×10^7 cells/ml

Cell preparations are kept on ice until transplantation experiments.

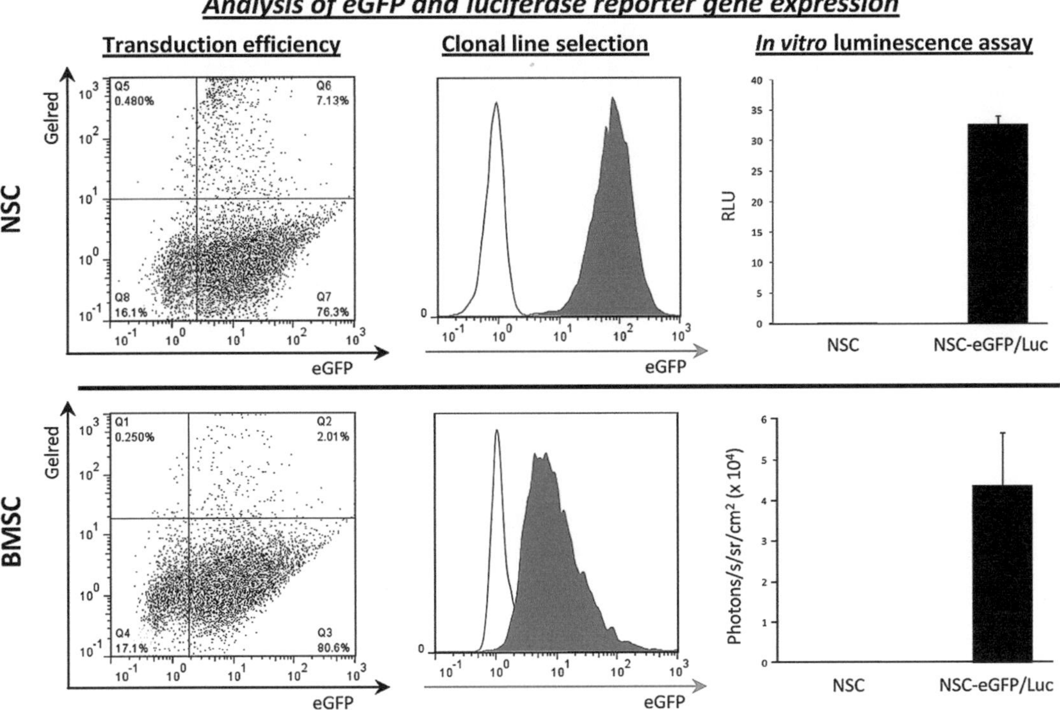

Fig. 2 Analysis of lentiviral vector transduction of NSC and BMSC with eGFP and Luciferase reporter genes. *Left dot plot figures*: Adherent NSC (*upper figure*) and BMSC (*lower figure*) cultures were transduced with a lentiviral vector expressing the eGFP–IRES–Luciferase cassette and were analyzed by flow cytometry for eGFP fluorescence (*x*-axis) versus viability (GelRed-staining, *y*-axis) after 48 h of culture. The percentage indicated in the *lower left quadrant* is the number of viable eGFP–Luc-positive cells, whereas the percentage in the *lower right quadrant* is the number of viable non-transduced (eGFP-negative) cells. The percentages indicated in the *upper left* and *right quadrant* are the numbers of nonviable cells. *Middle histogram overlays*: Representative flow cytometric analysis of eGFP expression of clonally expanded adherent NSC lines (*upper figure*) and BMSC lines (*lower figure*). *Open histogram*: control background fluorescence from parental NSC/BMSC. *Filled histogram*: direct eGFP fluorescence from clonal NSC-eGFP/Luc and BMSC-eGFP/Luc lines. *Right graphs*: In vitro luminescence assay on 1×10^5 parental and clonally expanded NSC-eGFP/Luc (*upper figure*) and BMSC-eGFP/Luc (*lower figure*). Data are expressed as relative light units or photons/s/sr/cm² from a 5-min time period ± standard error of two independent measurements

3.3.2 Cell Transplantation of eGFP–Luc-Expressing NSC and BMSC

Cell implantation was reproducibly targeted in the right hemisphere at following coordinates relative to bregma: 0 mm anterior, 2 mm lateral, and 2.5 mm ventral.

1. Anaesthetize the mice by an intraperitoneal injection of a ketamin/xylazin mixture and place in a stereotactic frame.
2. Shave the mice head and disinfect the skin.
3. Wet the eyes to prevent dehydration.
4. Make a midline scalp incision to expose the skull.
5. Drill a hole in the skull using a dental drill burr at bregma at 2 mm on the right side of the midline.
6. Vortex the cell suspension briefly and aspirate the cell suspension in a 10 µl Hamilton Syringe.
7. Place an automatic microinjector pump (kdScientific) with the syringe above the exposed dura.
8. Stereotactically place a 30-gauge needle (Hamilton) attached to the syringe through the intact dura to a depth of 2 mm.
9. Wait 1 min to allow pressure equilibration.
10. Inject 2×10^5 eGFP–Luc-expressing NSC or BMSC (2 µl) at a speed of 0.7 µl/min using an automatic microinjector pump.
11. Slowly retract the needle after 3 min to allow pressure equilibration and to prevent backflow of the injected cell suspension.
12. Disinfect the skin borders and suture the skin.
13. Administer a 0.9 % NaCl solution subcutaneously in order to prevent dehydration and place mice under a heating lamp to recover.

3.4 In Vivo Cell Graft Survival Analysis of Intracranially Implanted eGFP–Luc-Expressing NSC and BMSC Using Quantitative Bioluminescence Imaging

In order to obtain temporal information regarding stem cell graft survival following intracerebral implantation, bioluminescence imaging was performed at different time-points post-injection.

3.4.1 Bioluminescence Imaging

1. Anaesthetize the mouse by inhalation of an isoflurane/oxygen mixture for 5 min until mice are asleep.
2. Place the mouse in an in vivo real-time φ-imager system and reduce isoflurane levels to 1.5 % during the measurement.
3. Intravenously inject 150 mg/kg body weight D-Luciferin into the tail vein.
4. Image the mouse for 5 min.
5. Remove the mouse from the Photon Imager and allow to recover.

At the end of every acquisition obtain a photographic image to which the bioluminescence image can be superimposed by the analysis software.

Fig. 3 In vivo bioluminescence imaging analysis of cell graft survival following intracerebral implantation of eGFP–Luciferase-expressing NSC and BMSC. *Left figures: Intracerebral implantation.* Cell implantation was reproducibly targeted in the right hemisphere at the coordinates relative to bregma: 0 mm anterior, 2 mm lateral, and 2.5 mm ventral. The *circle* indicates the cell injection spot. *Right figures: in vivo bioluminescence imaging—image acquisition.* Representative time course images showing in vivo BLI of mice grafted with NSC-eGFP/Luc (*upper figures*) and BMSC-eGFP/Luc (*lower figures*) in the CNS. Images were acquired at day 1, day 3, day 7, and day 14 post-implantation. Light emission was measured from fixed regions of interest on the mouse head where NSC-eGFP/Luc and BMSC-eGFP/Luc were injected and on the mouse shoulder, which is considered as background signal. Values of signal intensity are presented as the average number of photons/s/sr/cm^2 from a 5-min time period

3.4.2 Quantitative Analysis of BLI Signals

Quantitative analysis of BLI signals was performed using M3 Vision software. Hereby, light emission was measured from a fixed region of interest on the mouse head, and values of signal intensity were presented as the average number of photons/s/sr/cm^2 over a 5-min time period. An additional region of interest was drawn on the mouse shoulder and considered as background signal.

3.4.3 Expected Outcome

Figure 3 provides representative results regarding in vivo BLI analysis of cell graft survival following intracerebral implantation of eGFP–Luciferase-expressing NSC and BMSC.

3.5 Investigating Cell Graft Survival and Endogenous Glial Cell Responses Following Intracerebral Stem Cell Transplantation via Quantitative Histological Analysis

1. Anaesthetize the mouse deeply by inhalation of an isofluorane/oxygen/nitrogen mixture for 2 min.

2. Sacrifice the mouse by perfusion with 4 % paraformaldehyde.

3. Remove the whole mouse brain from the skull and fixate in 4 % paraformaldehyde for 1 h.

4. Dehydrate the brain tissue by placing it subsequently in different gradients of sucrose (2 h in 5 %, 4 h in 10 % and overnight in 20 %).

5. Snap freeze the brain tissue in liquid nitrogen and store at −80 °C until sectioning.

3.5.1 Brain Dissection for Histological Analysis

3.5.2 Preparation of Tissue Slides for Histological Analysis

1. Section the entire implant region in serial 10 μm thick cryosections using a cryostat, hereby noting consecutively marked and missing slides.

2. Screen unstained cryosections directly using a fluorescence microscope to locate eGFP-expressing NSC and BMSC.

3. Store slides at −20 °C until histological analysis.

3.5.3 Histological Analysis: Determination of Cell Graft Volume

1. Obtain an image from every fifth slide of the obtained consecutive brain slides containing eGFP-expressing cells using an immunofluorescence microscope.

2. Determine the surface are (in XY plane provided as pixels^2) of the NSC-Luc/eGFP or BMSC-Luc/eGFP graft based on eGFP expression and converted to mm^2 (ImageJ).

3. Linearly extrapolate data from missing slides and slides in between based on acquired data.

This will allow the calculation of the total graft site volume, provided as graft site volume in mm^3 [=Sum of each individual graft volume per slide (=Graft site surface in XY plane in mm^2 × 10 μm)].

3.5.4 Histological Analysis: Immunofluorescence Staining

Histological analysis was performed to determine the presence of endogenous microglia (Iba-1) and astrocytes/astrogliosis (GFAP) according to the following procedure. All experimental procedures are performed at room temperature in the dark unless stated otherwise.

1. Use a DAKO pen to delineate the tissue section.

2. Rinse the section for 5 min with TBS.

3. For intracellular staining, permeabilize the section by incubating for 30 min in 0.1 % Triton-X.

4. Incubate the section for 1 h with 20 % blocking serum.

5. Incubate the section with primary antibody at appropriate dilution in antibody dilution buffer overnight at 4 °C.

6. The following day rinse 3 × 5 min with TBS.

7. Incubate the section with secondary antibody at the appropriate dilution in antibody dilution buffer for 1 h.

8. Rinse 3 × 5 min with TBS.

9. Counterstain the section with a nuclear stain by incubating the section for 20 min.

10. Rinse 2 × 5 min with TBS, followed by rinsing 2 × 5 min with distilled water.

11. Mount the slides with Prolong Gold antifade reagent and visualize immediately with a fluorescence microscope.

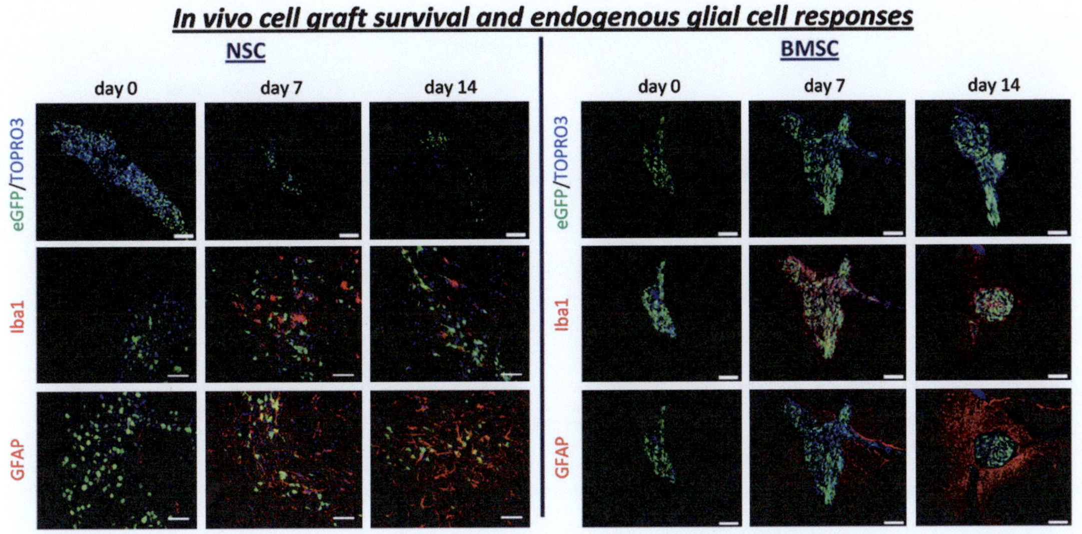

Fig. 4 Histological analysis of cell graft appearance and endogenous microglial and astroglial cell responses. *First row: cell graft survival.* Direct eGFP fluorescence (in *green*) combined with TOPRO3 staining (false color representation in *blue*). *Second row: microglial response.* Direct eGFP fluorescence (in *green*) combined with TOPRO3 staining (false color representation in *blue*) combined with immunofluorescence staining for Iba1 (in *red*). *Third row: astroglial response.* Direct eGFP fluorescence (in *green*) combined with TOPRO3 staining (false color representation in *blue*) combined with immunofluorescence staining for GFAP (in *red*). Representative images are chosen from multiple stained slides per mouse analyzed at day 0, day 7, and day 14 post-implantation. Scale bars: 50 µm and 200 µm

Figure 4 provides representative results regarding histological analysis of cell graft appearance and endogenous microglial and astroglial cell responses.

3.5.5 Quantitative Analysis of Immunofluorescence Stainings

Quantitative analysis of endogenous glial cell responses was performed using Image analysis software (TissueQuest, Photoshop, and ImageJ) according to the following procedure.

1. On each of the stained tissue slices manually delineate the following regions: (a) the implant site based on eGFP expression, and (b) an implant border 100 µm extending from the implant site (Adobe Photoshop).

2. For each of the delineated regions determine the surface area in XY plane (ImageJ).

3. Within these two regions acquire the following data for each mouse analyzed: (a) the number of eGFP-expressing grafted NSC, (b) the number of Iba1[pos] microglia (TissueQuest) and the number of S100B[pos] or GFAP[pos] astrocytes (TissueQuest).

Figure 5 provides a general overview regarding the procedure for quantitative analysis of immunofluorescent images with Tissue-Quest Software.

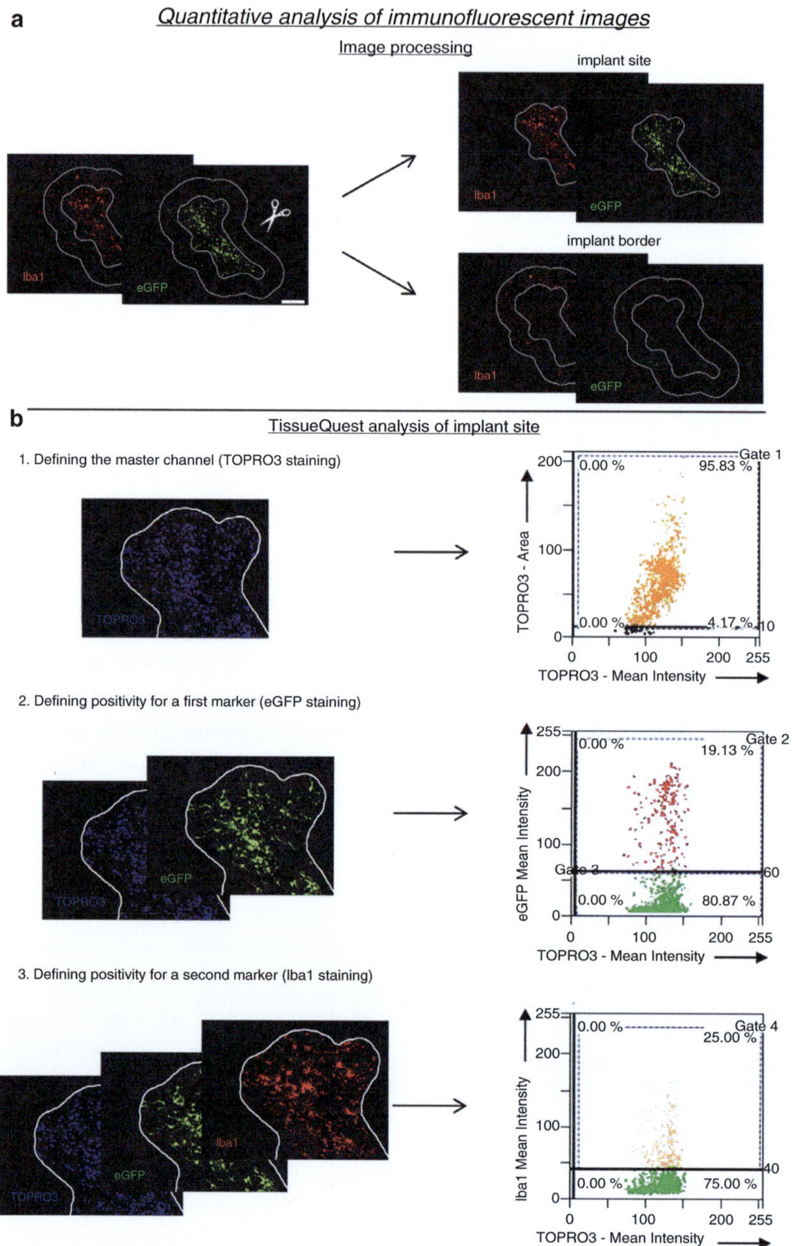

Fig. 5 Quantitative analysis of immunofluorescent images with TissueQuest Software. (**a**) *Image processing*: Use image analysis software to delineate (1) the implant site based on eGFP expression, and (2) an implant border 100 μm extending from the implant site for each of the color images per stained tissue slide. (**b**) *TissueQuest analysis of implant site*: First, define the master channel based on nuclear staining with TOPRO3, using TissueQuest analysis software. Gate for nuclei based on TOPRO3 positivity and size (Gate 1). Second, define positivity for eGFP on the gated (Gate 1) population, based on eGFP intensity. This will allow the identification of positive (Gate 2) and negative (Gate 3) eGFP-expressing cells within the implant site. Third, define positivity for Iba1 within the endogenous eGFP-negative cell population (Gate 3) based on Iba1 intensity. Finally, note the total number of cells within the gated populations as provided by TissueQuest analysis software

For data presentation: Calculate the cellular density of each population based on the known surface area in the XY plane of the analyzed slides for each of the two regions per mouse brain as follows: mean of # cell counts for a specific population/[(surface area in mm^2) $\times$ 10 μm] and provided per mouse as: (a) # eGFP-expressing grafted NSC/mm^3, (b) # Iba1[pos] microglia/mm^3, and (c) # S100B/GFAP[pos] astrocytes/mm^3.

References

1. Reekmans K, Praet J, Daans J et al (2012) Current challenges for the advancement of neural stem cell biology and transplantation research. Stem Cell Rev 8:262–278

2. Reekmans K, Praet J, De Vocht N et al (2012) Stem cell therapy for multiple sclerosis: preclinical evidence beyond all doubt? Regen Med 7:245–259

3. Ronsyn MW, Berneman ZN, Van Tendeloo VF et al (2008) Can cell therapy heal a spinal cord injury? Spinal Cord 46:532–539

4. Bergwerf I, De Vocht N, Tambuyzer B et al (2009) Reporter gene-expressing bone marrow-derived stromal cells are immune-tolerated following implantation in the central nervous system of syngeneic immunocompetent mice. BMC Biotechnol 9:1

5. Reekmans KP, Praet J, De Vocht N et al (2011) Clinical potential of intravenous neural stem cell delivery for treatment of neuroinflammatory disease in mice? Cell Transplant 20:851–869

6. Ronsyn MW, Daans J, Spaepen G et al (2007) Plasmid-based genetic modification of human bone marrow-derived stromal cells: analysis of cell survival and transgene expression after transplantation in rat spinal cord. BMC Biotechnol 7:90

7. Tambuyzer BR, Bergwerf I, De Vocht N et al (2009) Allogeneic stromal cell implantation in brain tissue leads to robust microglial activation. Immunol Cell Biol 87:267–273

8. Bergwerf I, Tambuyzer B, De Vocht N et al (2011) Recognition of cellular implants by the brain's innate immune system. Immunol Cell Biol 89:511–516

9. De Vocht N, Lin D, Praet J et al (2013) Quantitative and phenotypic analysis of mesenchymal stromal cell graft survival and recognition by microglia and astrocytes in mouse brain. Immunobiology 218(5):696–705

10. Praet J, Reekmans K, Lin D et al (2012) Cell type-associated differences in migration, survival and immunogenicity following grafting in CNS tissue. Cell Transplant 21 (9):1867–1881

11. Tambuyzer BR, Ponsaerts P, Nouwen EJ (2009) Microglia: gatekeepers of central nervous system immunology. J Leukoc Biol 85:352–370

12. De Vocht N, Bergwerf I, Vanhoutte G et al (2011) Labeling of Luciferase/eGFP-expressing bone marrow-derived stromal cells with fluorescent micron-sized iron oxide particles improves quantitative and qualitative multimodal imaging of cellular grafts in vivo. Mol Imaging Biol 13:1133–1145

13. De Vocht N, Reekmans K, Bergwerf I et al (2012) Multimodal imaging of stem cell implantation in the central nervous system of mice. J Vis Exp 13:e3906

Methods Molecular Biology (2013) 1052: 143–152
DOI 10.1007/7651_2013_23
© Springer Science+Business Media New York 2013
Published online: 3 May 2013

Micro-CT Technique for Three-Dimensional Visualization of Human Stem Cells

Andrea Farini, Chiara Villa, Marzia Belicchi, Mirella Meregalli, and Yvan Torrente

Abstract

Micro-CT offers high spatial resolution of the distribution of stem cells and provides rapid reconstruction of 3D images and quantitative volumetric analysis. Together with real-time PCR analysis, micro-CT offers the possibility to obtain a quantification of the number of cells that are able to migrate from the bloodstream inside the muscular tissues. Here, we studied for the first time the kinetics of the human cells injected into the femoral artery of DMD animal model. It is fundamental to determine whether the cells disseminate and entrap only within the capillary system of downstream muscles and/or they are able to reach the non-injected muscles and other organs through blood flow. The efficient transplantation of stem cells to dystrophic-deficient muscle reinforced the utility of intra-arterial delivery of cells as a viable approach for cell-based clinical therapies of neuromuscular diseases.

Keywords: Micro-CT, 3D images, CD133+ stem cells, Muscle homing, Muscular dystrophy

---

1 Introduction

Cell therapy is an emerging area of research in regenerative medicine with significant efforts being carried out in several clinically important areas (1). Since the feasibility of the stem cells for cellular therapies was determined, another important point was their administration route: intravenous administration of stem cells had the most immediate access to clinical applications in patients suffering from stroke and it was used in experimental studies with variable success (2). In a clinical perspective to treat devastating diseases as muscular dystrophies, it is necessary that the stem cells could reach the whole body, as all the muscles of the patient suffered for the pathology. Consequently, vascular architecture, expression of specific molecules, discovery of mechanisms involved in muscle homing of stem cells, and demonstration that various signals are thought to augment the recruitment of progenitor cells into the damaged tissue for repair improved a potential therapy for a large range of diseases on the systemic delivery of such cells (3, 4). The ability to noninvasively monitor cell trafficking in vivo in a longitudinal fashion is a pressing need for emerging cellular therapeutic strategies. Ideally, imaging technology used for stem cell tracking should have single-cell sensitivity, permit quantification of exact cell numbers at any anatomic location, and determine the transplanted cells' engraftment efficiency and functional capability. Micro-CT is similar to conventional CT systems usually employed in medical diagnoses and industrial applied research (5–7). The principle of micro-CT is the use of the attenuation of X-rays by various tissues spaced within angular intervals. The micro-CT offers high spatial resolution of the distribution of stem cells, provides rapid reconstruction of 3D images and quantitative volumetric analysis, and, most importantly, does not alter the labeled biological process itself (8, 9). In this work, we demonstrated that the micro-CT imaging could be applied to show the distribution of intra-arterially administered super paramagnetic nanoparticle-labeled CD133+ stem cells within muscle biopsies, providing biological insights into the early processes of muscle stem cell homing.

2   Materials

Prepare all solutions using deionized water and analytical grade reagents. Prepare and store all reagents at room temperature (RT) (unless indicated otherwise). Diligently follow all waste disposal regulations when disposing waste materials.

2.1 Isolation of Human CD133+ Cells from Blood

1. RPMI 1640 medium (Invitrogen Life Technologies, Grand Island, NY, USA).

2. Ficoll-Hystopaque 1.077 g/mL (Sigma & Aldrich, St. Louis, Missouri, USA).

3. Midi-MACS and LS columns (Miltenyi Biotec, Bergisch Gladbach, Germany).

4. PBS 1×: Add about 100 mL water to a 1-L graduated cylinder (or a glass beaker). Weigh 4 g NaCl, 0.1 g KCl, 0.72 g Na_2HPO_4, and 0.12 g KH_2PO_4 and transfer to the cylinder. Add water to a volume of 700 mL. Mix and adjust pH until it reaches 7.4. Make up to 1 L with water. Store at 4 °C.

5. EDTA: Weight 0.75 g of ethylenediaminetetraacetic acid, put in beaker, and dissolve in 100 mL of sterile water at pH 7.4, by using a stirring magnetic bar (see Note 1).

6. Suspension buffer: Add 20 mL of EDTA (20 mM) to a 250 mL glass beaker. Weigh 1 g bovine serine albumin (BSA), transfer to the beaker, and add PBS 1× to reach the total volume of 200 mL. Filter and store at 4 °C (see Note 2).

7. Monoclonal antibody (clone CD133, epitope1) (Microbead Kit, Miltenyi).

8. 30 µm pre-separation filters (Miltenyi).

9. Proliferation medium (PM) composed of DMEM/F-12 (Euroclone, Pero, Italy) (1:1); 20 % FBS (Euroclone), HEPES buffer (5 mM, Gibco-Invitrogen, Carlsbad, USA), glucose (0.6 %), sodium bicarbonate (3 mM), and glutamine (2 mM); SCF (100 ng/mL) (TEBU, Frankfurt, Germany), VEGF (50 ng/mL) (TEBU), and LIF (20 ng/mL) (R&D Systems Inc, Minneapolis, Minnesota, USA).

10. Antibodies for FACS: Anti-CD133-phycoerythrin (anti- CD133-PE), anti-CD34-APC, anti-CD31-FITC, anti-CD45-APC-CY7, anti-7-amino-actinomycin D (anti-7AAD), anti-CD44 FITC, anti-CXCR4 FITC, anti-CD146PE (BD Biosciences-Pharmigen, San Diego, California, USA); anti-CD90 PE-Cy5, anti-CD146 PE, anti-CD105 PE, anti-CD117 PE (Beckman Coulter, Brea, California, USA).

11. 2 mL not sterile polystyrene tubes (Beckman Coulter).

12. Cytomic FC 500 flow cytometer and CXP 2.1 software (Beckman Coulter).

13. 48-well plates (Corning Inc, Corning, New York, USA).

2.2 Labeling of CD133 + Cells with Endorem

1. Endorem (250 µg/mL nanoparticles of FeO) (Guerbet Roissy, CdG Cedex, France).

2. Prussian Blue: Add 40 mL of HCl (0.06 N) to a glass beaker. Weigh 0.4 g of potassium hexacyano-ferrate (II) trihydrate and gently mix.

2.3 In Vivo Transplantation

1. Prepare 100 % Avertin stock solution by dissolving 5 g of 2,2,2-tribromoethanol in 5 mL of 2-methyl-2-butanol. Mix for 4 h or until fully dissolved. Prepare 2.5 % working solution by adding 5 mL of the above solution dropwise on 195 mL of 0.15 M saline in a 200 mL screw cap bottle and filter through a 0.2 µm Sterile Syringe Filter (Corning). Store at 4 °C in a conical tube covered with aluminum foil (see Note 3).

2. 1 mL sterile, pyrogen-free syringe (30 G needle).

3. Stereotaxic microscope.

4. Small sterile surgery scissors (see Note 4).

5. Two pairs of curved forceps.

6. Gauzes to clean surgery instruments.

7. Some suture string with curved needle (size 3-0, hooked 26 mm needle).

8. Empty cages.

9. Heating lamp.

2.4 Micro-CT Procedure	1. ID19 beamline of the European Synchrotron radiation facility (ESRF) in Grenoble—France.
	2. Liquid nitrogen.
	3. Liquid nitrogen-cooled isopentane.
	4. Tweezers.
	5. Cryovials.

2.5 Image Reconstruction

1. VG-Studio Max 1.2 software.

2.6 Absolute Real-Time PCR (Q-PCR)

1. Trizol Reagent (Invitrogen).

2. DEPC water: To reduce the risk of RNA being degraded by RNases, treat 1 L of sterile water with 0.1 % v/v diethylpyrocarbonate, stirring overnight at RT in glass bottle, and then autoclave to inactivate traces of DEPC.

3. BioPhotometer (Eppendorf, Hamburg, Germany).

4. DNAse–RNAse free (Promega, Madison, WI, USA).

5. Super Script First Strand III Synthesis System for RT-PCR (Invitrogen).

6. SYBR Green (EuroClone).

7. Chromo 4, Real-Time PCR detector, PTC 200 (MJ Research BioRad, Hercules, CA, USA).

8. Human GAPDH primers F: 5′-GTGGCAAAGTGGAGATTGT TGCC-3′; R: 5′-GTAGATGACCCGTTTGGCTCC-3′.

3 Methods

Carry out all procedures at RT unless otherwise specified.

3.1 Isolation of Human CD133+ Cells from Blood

1. Collect blood CD133+ cells from peripheral blood of normal volunteers after informed consent, according to the guidelines of the Committee on the Use of Human Subjects in Research at the Fondazione IRCCS Cà Granda OMP (Milan, Italy).

2. Form each blood sample by a pool of four samples collected from 18- to 50-year-old healthy volunteers. Dilute 100 mL of blood 1:1 in RPMI 1640 medium (Invitrogen) and centrifuge for 10 min at 800 rpm to remove the platelet-rich plasma phase.

3. Use warm Ficoll-Hystopaque 1.077 g/mL (Sigma Aldrich) density centrifugation to obtain mononuclear cells (MNC).

4. Carefully collect the MNC ring between plasma phase (above) and Ficoll (under).

5. Wash the mononuclear cells obtained in suspension buffer.

6. Isolate the CD133+ MNC fraction by using microbead selection, MidiMACS and LS columns: Suspend the MNC cells in PBS–0.1 % BSA and 2 mM EDTA buffer and directly label cells by incubation for 30 min at 4 °C with a monoclonal antibody clone CD133, epitope1, coupled with microbeads (see Note 5).

7. Wash the cells in cold suspension buffer, resuspend up to 10^8 cells in 500 μL of buffer, and lay on an LS column placed in a magnetic field. Use 30 μm pre-separation filters, to eliminate clusters which may clog the column. The magnetically labeled CD133+ cells retain on the column, while unlabeled CD133+ cells pass through. Remove the column from the magnetic field and elute magnetically retained CD133+ cells as a positively selected fraction (see Notes 6 and 7).

8. In order to obtain a higher purity, repeat the previous procedure (step 5) by passing through the column the eluted CD133+ cells (see Note 8).

9. Count living cells with a Burker chamber using Trypan Blue exclusion: Resuspend the cells into Trypan Blue 0.2 % (diluted 1:1) (Sigma Aldrich). After few minutes, count the cells (death cells will appear blue-colored).

10. Analyze CD133+ cell purity and vitality through six-color flow cytometry: Collect 2×10^4 cells in 2 mL non-sterile polystyrene tube and incubate with FACS antibodies anti-CD133-phycoery-thrin (anti-CD133-PE, Miltenyi) and anti-7-amino-actinomycin D (anti-7AAD) (BD Biosciences-Pharmigen). For the control samples, add isotype-matched mouse immunoglobulins. Incubate each sample at 4 °C for 20 min. Wash cells to stop the reaction in suspension buffer, centrifuge at $458 \times g$ for 10 min, and resuspend the pellet in 300 μL of suspension buffer, for the FACS analysis. Each analysis included at least 10,000–20,000 events for each gate. Eliminate cell debris from the analysis by using a light-scatter gate: the percentage of positive cells is assessed after correction for the percentage reactive to an appropriate isotype-matched mouse immunoglobulin, used as a control.

11. Repeat the same procedure of step 8 to characterize cells by FACS analysis with antibodies anti-CD90 PE-Cy5, anti-CD105 PE (Beckman Coulter), anti-CD34-APC, anti-CD31-FITC, anti-CD45-APC-CY7, anti-CD146 PE, anti-CD44 FITC, anti-CXCR4 FITC, and anti-CD117 PE (BD Biosciences-Pharmigen) (see Note 9).

12. Seed the cells in 48-well plates in a PM at a density of 15×10^4 cells/cm^2.

***3.2 Labeling of CD133
+ Cells with Endorem***

1. Label cells with 250 µg/mL nanoparticles of Endorem: Add 15.8 µL of Endorem to 1 mL of RPMI 1640 medium enriched with 20 ng/mL epidermal growth factor (EGF) and 10 ng/mL basic fibroblastic growth factor (bFGF) and incubate at 37 °C for 24 h. Wash cells, resuspending them in PBS 1×, and centrifugate at 3,000 × g for 2 min. Repeat this step three times.

2. In order to perform Prussian blue staining, to visualize cells, add in 48-well plates 200 µL of C_6FeKN_6 solution for 1 h at RT. Wash cells, resuspending in tap water and centrifuging at 3,000 × g for 2 min. Repeat this step three times.

3. Visualize cells as blue spots by using Leica ALSD microscope (Leica, Hessen Wetzlar, Germany).

***3.3 In Vivo
Transplantation***

1. Anesthetize the mouse by injecting 300 µL of Avertin 2.5 % intraperitoneally (for a mouse weighting ~20 g).

2. Check the reflex by pinching the finger.

3. Flip the mouse on its back, maintained by some rubbers on the limbs.

4. Put the mouse under the stereotaxic microscope to proceed with the injection.

5. Wet the skin of the right leg with ethanol 70 %. Eliminate the hairs and make a vertical incision into the skin with scissors. The incision should be 0.8–1.2 cm, starting 1 cm below the groin.

6. Stretch the borders with the forceps to see the femoral artery.

7. Inject the cells in the artery by means of the 1 mL syringe, being careful not to touch the other vessels on both sides.

8. Close the skin with at least two clips.

9. Put the mice in a clean cage and warm it until they are recovered.

***3.4 Micro-CT
Procedure***

1. For in vivo scans, put anesthetized transplanted mice with right leg fixed in a cylindrical plastic holder so that the thigh region is exposed to the X-ray beam.

2. Set the X-ray beam energy to 24 keV. The total radiation dose on mouse is 40–45 Gy.

3. As a detection system, use a Gadox scintillator associated to a FReLoN 2,048 × 2,048 pixel CCD camera, with the pixel size set to 7.05 µm. The field of view measures 14.4 × 14.4 × 1.8 mm^3 (2,048 × 2,048 × 256 voxels).

4. Analyze 700 projections.

5. Set the acquisition time to 0.2 s/projection.

6. Take tomograms of the thigh region at 0, 2, 13, and 24 h after injection in three different consecutive regions of interest along the femur direction, for a total thickness of 5.4 mm.

7. Sacrifice animal at the end of micro-CT acquisition by cervical dislocation: All procedures involving living animals were conformed to Italian Country law (D.L.vo 116/92).

8. Remove muscle tissues and organs and put in liquid nitrogen-cooled isopentane for 30 s by means of tweezers (see Note 10).

9. Put muscle tissues and organs in liquid nitrogen for 1 min and store in cryovials at $-80\ ^\circ$C.

3.5 Image Reconstruction

1. Reconstruct isotropic slice data into 2D images.

2. Combine the multiple projections using a reconstruction method based on the filtered back projection implemented with a projection algorithm. The resultant micro-CT scan is a 3D matrix of voxels with values proportional to the mean linear attenuation coefficient of the material within each voxel. The image obtained on the CCD-based detector depends on the linear absorbing coefficients of the phases/materials that compose the investigated specimen (see Note 11).

3. Once the reconstruction of the field of view area is done using specific algorithms, analyze the obtained volume using the VG-Studio Max 1.2 software that allows visualizing and quantifying the different phases/materials in the volume (see Note 12).

4. Using the segmentation facility in the VG-Studio Max software, quantify the fat, muscle, bone, and cells present in the volume. As it is experimentally impossible to view exactly the same volume for the four measurements (at 0, 2, 13, and 24 h), correct the position by quantifying the total fat + muscle volume for each of the four measurements and rescaling it (horizontally) until an overlap was achieved (see Notes 13 and 14).

5. The investigated area of the leg was vertically divided in 1,000 slices, for the reconstruction with 0 corresponding to the upper part of the field of view and 1,000 to the lower part of it.

3.6 Absolute Real-Time PCR (Q-PCR)

1. Extract total RNA with Trizol Reagent as described by the manufacturer's protocol (Invitrogen) from frozen muscle tissues and organs (see Note 15).

2. Measure the absorbance of each sample by means of the Bio-Photometer: Mix in the cuvette 1 μL of RNA sample with 49 μL of DEPC water and calculate the concentration of RNA (expressed as μg/μL): the ratio Δ260/280 should be comprised between 1.8 and 2 (see Note 16).

3. Treat 2 μg of each RNA sample with RQ1 RNase-Free DNase to avoid amplification of genomic DNA as described by the manufacturer's protocol (Promega).

4. Prepare first-strand cDNA with oligo(dT)$_{12-18}$ primers as described by the manufacturer's protocol (Invitrogen) (total volume: 50 μL).

5. As Q-PCR was used to obtain an absolute quantification of human GAPDH in the muscles and organs of injected mice, prepare a calibration curve: 2 ng of plasmid carrying the sequence of human GAPDH were serially diluted (1:10) eight times.

6. Prepare each sample by taking 1.5 μL of cDNA, mix with MasterMix PCR for SYBR Green (FluoCycle, Euroclone) 1× (10 μL), and forward and reverse primers (3 μM) (see Section 2.6, item 2) (2 μL each), and reach a final volume of 20 μL (see Note 17).

7. Run the program: 50 cycles, annealing temperature 60 °C, melting curve of 30 min to verify the specificity of the primers (see Note 18).

8. Measure GAPDH levels in Q-PCR by SYBR GREEN technique and calculate the picograms of human–GAPDH per muscle/organ per grams of muscle/organ (see Note 19).

4 Notes

1. Measure carefully the pH as the EDTA will dissolve only when the pH is comprised between 7 and 8.

2. BSA can be replaced by other proteins such as human serum albumin (HSA), human serum, or fetal bovine serum (FBS).

3. It is possible to dissolve the 2,2,2-tribromoethanol in PBS 1×.

4. To sterilize the scissors use glass bead sterilizer.

5. During the incubation at +4 °C, gently shake the tube, to avoid cell pellet deposition.

6. Prepare the column by rinsing with 3 mL of buffer and then apply MNC into the column.

7. Resuspend mononuclear cellular pellet in 300 μL of buffer per 10^8 total cells. Add 100 μL of FCR blocking reagent per 10^8 total cells. Add 100 μL of CD133 MicroBeads per 10^8 total cells. Mix well and incubate for 30 min in the refrigerator (2–8 °C).

8. It is recommended not to use 30 μm pre-separation filter, to avoid losing cells during the second elution through the column.

9. Steps 8 and 9 can be performed in the same FACS analysis, depending on the cell number obtained after column.

10. Use tweezers at RT. Frozen tweezers could ruin the tissue. For prolonged use, put some water into a glass backer, warm up through a plate, and leave the tweezers inside.

11. The labeled cells are visualized as red spots in the reconstructed 3D volumes. A more accurate analysis of the spatial distribution of the nanoparticles can be achieved by deleting the other tissue phases by software.

12. The volume fractions of migrated labeled cells are calculated by counting their corresponding pixels.

13. The small movements of the anesthetized animals, essentially due to the heartbeat with a frequency of 10 Hz, induce two oscillations of the tissue during the acquisition time of each projection. Therefore the beam detects an apparent cell size comparable to twice the oscillation amplitude (200–300 μm). On this basis, you estimate a factor ~50 in the measured volume fraction with respect to the real one.

14. Put into evidence labeled cell eliminating by software all the other tissues in the image; you cannot cancel only the femur bone because of its absorption coefficient similar to the Endorem one and the two peaks in the grey level histogram are overlapped.

15. During RNA extraction, if it would be impossible to resuspend the pellet, heat it at 65 °C for 2 or 3 min.

16. Vortex briefly the nucleic acid sample prior to dilution to avoid possible concentration gradient in the sample. Before transferring the sample into the cuvette vortex briefly again to prevent concentration fluctuations, usually caused by long storage of the sample.

17. Depending on the available real-time instrument (see Section 2.6, item 2), it could be possible to use Real Master SYBR ROX Mix (EuroClone).

18. For SYBR Green-based amplicon detection, it is necessary to run a dissociation curve following the real-time PCR cycles. In fact SYBR Green detects any double-stranded DNA including primer dimers, contaminating DNA, and PCR product from mis-annealed primer. By viewing a dissociation curve, you ensure that the desired amplicon was detected.

19. To obtain the correct amount of picograms of human–GAPDH per muscle/organ per grams of muscle/organ it is fundamental to identify the cycle number at which the increase in fluorescence is exponential. It is called threshold line and is set by the user: in our experiment we set the line at 0.1 level of fluorescence.

Acknowledgments

Stem Cell Laboratory, Dipartimento di Fisiopatologia Medico-Chirurgica e dei Trapianti, Università degli Studi di Milano, Fondazione IRCCS Cà Granda OMP Milano was supported by Associazione La Nostra Famiglia Fondo DMD Gli Amici di Emanuele, Associazione Amici del Centro Dino Ferrari, and EU's 7th Framework programme Optistem 223098.

References

1. Teo AK, Vallier L (2010) Emerging use of stem cells in regenerative medicine. Biochem J 428:11–23
2. Guzman R, Choi R et al (2008) Intravascular cell replacement therapy for stroke. Neurosurg Focus 24:E15
3. Bachrach E, Perez AL et al (2006) Muscle engraftment of myogenic progenitor cells following intraarterial transplantation. Muscle Nerve 34:44–52
4. Schulz C, von Andrian UH, Massberg S (2009) Hematopoietic stem and progenitor cells: their mobilization and homing to bone marrow and peripheral tissue. Immunol Res 44:160–168
5. Namati E, Chon D et al (2006) In vivo micro-CT lung imaging via a computer-controlled intermittent iso-pressure breath hold (IIBH) technique. Phys Med Biol 51:6061–6075
6. De Clerck NM, Meurrens K et al (2004) High-resolution X-ray microtomography for the detection of lung tumors in living mice. Neoplasia 6:374–379
7. Marechal M, Luyten F et al (2005) Histomorphometry and micro-computed tomography of bone augmentation under a titanium membrane. Clin Oral Implants Res 16:708–714
8. Badea CT, Drangova M et al (2008) In vivo small-animal imaging using micro-CT and digital subtraction angiography. Phys Med Biol 53:R319–R350
9. Torrente Y, Gavina M et al (2006) High-resolution X-ray microtomography for three-dimensional visualization of human stem cell muscle homing. FEBS Lett 580:5759–5764

Methods Molecular Biology (2013) 1052: 153–166
DOI 10.1007/7651_2013_14
© Springer Science+Business Media New York 2013
Published online: 4 June 2013

Noninvasive Multimodal Imaging of Stem Cell Transplants in the Brain Using Bioluminescence Imaging and Magnetic Resonance Imaging

Annette Tennstaedt, Markus Aswendt, Joanna Adamczak, and Mathias Hoehn

Abstract

Transplantation of stem cells represents a promising approach for the therapy of different brain diseases, including stroke, Parkinson's, and Huntington's disease. Tracking of stem cells with noninvasive imaging technologies provides insight into location, migration, and proliferation of the cells—key features for a possible clinical translation. This chapter describes a multimodal and noninvasive approach employing magnetic resonance imaging (MRI) and bioluminescence imaging (BLI), both of which offer the opportunity for repetitive measurements on the same individual, revealing the full temporal profile of cell dynamics. The combination of these modalities allows the simultaneous investigation of different aspects of the graft fate. We will present the detailed protocol for noninvasive multimodal tracking of labeled and transplanted neural stem cells, specifically optimized for brain applications, which allows repetitive assessment of localization as well as identification of cell viability and cell quantity after transplantation.

Keywords: Magnetic resonance imaging (MRI), Bioluminescence imaging (BLI), SPIO nanoparticles, PrestoBlue assay, Neural stem cells (NSCs), Cell labeling, Intracerebral transplantation

1 Introduction

The transplantation of stem cells is intensively investigated as a novel therapeutic strategy for many diseases, with strong focus on neurodegenerative diseases. At present, conventional approaches require histological evaluation to assess the location and fate of grafted cells. The development and optimization of transplantation therapies therefore highly benefits from noninvasive imaging strategies, which allow repetitive assessment of cell behavior over chronic time periods. Noninvasive imaging methods are aimed to determine transplantation efficiency and to monitor therapy progress and outcome, prior to a clinical translation. Multimodal imaging approaches have the power to provide complementary information, i.e., cell localization, viability, and differentiation status (1). The investigation of the therapeutic potential of stem cells for neurodegenerative diseases imposes additional difficulties for noninvasive imaging. Knowledge from systemic transplantations (1, 2) has to be adapted for applications in the brain (3). MRI and Bioluminescence imaging (BLI) are both noninvasive techniques, which can be combined for multimodal imaging of stem cell grafts, and both have a broad application potential for brain imaging.

MRI offers the key advantage of a superior spatial resolution, which can reach 50 μm or even below, allowing even single cell detection (4) under ideal conditions. Furthermore, MRI provides soft tissue contrast, to discriminate e.g., between gray and white matter in the brain. For stem cell tracking, contrast is achieved by labeling the cells with MRI contrast agents in order to enable the discrimination of transplanted cells from the host tissue. MRI contrast agents modify the relaxation times of the surrounding protons, represented by changes in T1 and T2/T2* relaxation times. These relaxation times are responsible for the contrast in common T1- and T2-weighted MR images (5–8). Cells are usually labeled in vitro before transplantation, but also in vivo labeling of special cell populations is possible and has been successfully performed (9–14). Superparamagnetic iron oxide nanoparticles are the most commonly used contrast agents for cell tracking, because of their potential to efficiently modulate the MR signal in their surrounding—visible with a T2- and T2*-weighted sequence.

BLI, although it has a very low spatial resolution compared to MRI, offers a much higher sensitivity so that even small numbers of cells can be detected (15). BLI is based on an enzymatic light production. The emitted light is subsequently captured by a highly sensitive CCD camera. The enzyme in action is a luciferase which uses the substrate luciferin to produce light. Different luciferases from different species exist (firefly, click beetle, jellyfish, bacteria, etc.) and the emission spectrum differs between the various luciferases (16). In some species, like the firefly, the enzyme reaction is ATP-dependent; thus its light reaction can serve as a viability marker. In order to employ bioluminescence for cell tracking, cells have to be transgenically modified to express luciferase. Then, cells can be tracked over long periods of time since luciferase expression is preserved during proliferation or differentiation so that all cell progeny express luciferase without dilution (17). After systemic injection of the substrate, photon emission takes only place at the site of luciferase expression without unspecific background signal. Furthermore, light emission is directly proportional to the number of cells, thus allowing cell quantification (15, 18). Due to the method's high detection sensitivity, signal from deep in the brain may still be detected (15, 19). However, for deep tissue embedded signal light scattering and absorption will reduce the signal intensity.

It was shown that predifferentiated stem cells are a better choice for transplantation studies compared to native embryonic stem cells (2), therefore a neuronal stem cell line is the cell of choice for brain transplantation. In the following, we will describe a protocol for multimodal imaging of luciferase positive (Luc+) neuronal stem cells labeled with iron oxide particles for tracking transplanted cells in the brain. We will outline in detail the labeling procedure and the assessment of potential label effects on cell viability. We will continue with the description of the transplantation procedure into the brain, and will provide details on the imaging protocols for in vitro BLI and MRI as well as protocols for repetitive in vivo BLI and MRI.

2   Materials

2.1   Cell Culture Components

1. Medium for NSCs: N2Euromed (Biozol, Siziano, Italia) with 1× N2-Supplement (Life technologies, Darmstadt, Germany), 2 mM L-Glutamin (PAA, Cölbe, Austria), 100 U/ml Penicillin, 100 μg/ml Streptomycin (PAA), 50 μg/ml BSA (PAA), 25 μg/ml Insulin human (Sigma Aldrich, Taufkirchen, Germany), 5 ng/ml EGF (Pepro Tech, Hamburg, Germany) and 10 ng/ml FGF-2 (Pepro Tech) (see Note 1).

2. Solution for detachment of cells: Accutase (PAA).

3. SPIO nanoparticles (stock solution 11.2 mg/Fe/ml; Endorem®, Guerbet, France).

4. PBS solution (PAA).

5. Buffer for cell suspension: HBSS buffer (Life technologies).

6. 21-cm^2 tissue culture dish (Greiner, Solingen, Germany).

7. 96-well plate (Greiner).

8. PrestBlue solution (10× stock solution, Life technologies).

9. Plate Reader (Berthold, Bad Wildbad, Germany).

2.2   Surgery Components

1. For anesthesia: Isoflurane (2 %) (Pfizer, Berlin, Germany).

2. Analgetic drug: Carprofen (4 mg/kg) (Pfizer).

3. Needle for cell implantation: Hamilton syringe 5 μl with 26 G needle (Catalog number 7634-01, 7732-06) (Hamilton, Höchst, Reno, USA).

4. Surgical instruments (scissors, forceps) (Mesculap, Tuttlingen, Germany).

5. Syringes (1 ml), needles (20 G) (Catalog number 916 6017V, 4657519) (Braun, Melsungen, Germany).

6. Suture (noncolored) (Catalog number PGA Resoquick 5/0 USP) (Resorba, Nürnberg, Germany).

7. Stereotactical instrument (Catalog number 51600) (Stoelting, Dublin, Ireland).

8. Temperature controller (Medres, Köln, Germany).

9. Electric Driller (Catalog number Technobox 810) (Bien-Air, Bienne, Switzerland).

10. Digital navigation system (Catalog number 51904) (Stoelting).

11. Digital injection controller (Catalog number Micro 4) (World Precision Instruments, Berlin, Germany).

12. Physiological saline (NaCl 0.9 %) (Braun).

2.3 BLI Components

1. For in vitro BLI prepare the D-luciferin stock solution (Synchem, Felsberg, Germany) by dissolving the powder in phosphate buffered saline (PBS) to reach a concentration of 10 mM, filter sterile and keep dark at -20 °C for long-term storage.

2. For in vivo BLI prepare the D-luciferin stock solution (Synchem) by dissolving the powder in PBS to reach a concentration of 20 mg/ml, filter sterile and keep dark at -20 °C for long-term storage.

3. Black 96-well plate (Greiner).

4. For anesthesia: Isoflurane (2 %) (Pfizer).

5. Syringes (1 ml), needles (20 G) for injection (Catalog number 916 6017 V, 4657519) (Braun).

6. Biospace Photon Imager (Biospace Lab, Paris, France).

2.4 MRI Components

1. High-resolution magnetic images were acquired on a Biospec 11.7 Tesla (T) animal scanner system (Bruker, Karlsruhe, Germany) with a 16-cm horizontal bore magnet, equipped with actively shielded gradient (750 mT/m) using ParaVision 5 software.

2. For phantom signal detection use a rat brain surface coil (Bruker).

3. For mouse brain signal detection use a mouse quadrature surface coil (Bruker).

4. RF transmission was achieved with a 9 cm diameter resonator (Bruker).

5. 200 µl tube (Biozym, Wien, Austria).

6. Before scanning the phantom prepare 1 % agarose: Dissolve 0.1 g agarose (Sigma Aldrich) respectively agarose type VII (Sigma Aldrich) into 10 ml PBS to get a 1 % agarose solution. Warm it up while stirring.

7. Mount the phantom on a rat holder (Bruker).

8. Before scanning the mouse prepare a 2 % agarose cap: Make a 2 % agarose gel by mixing 0.2 g of agar into 10 ml of PBS.

Warm it up while stirring and pour it into a moon shaped container (see Note 2). Let it cool down and get solid for 20 min in a fridge.

9. Mount the mouse on a mouse holder (Bruker) using tooth bar and ear bars for stable positioning.

10. Bepanthen® cream (Bayer, Leverkusen, Germany).

11. Control and monitor body core temperature with an in-house designed automated temperature control unit complete with water blanket.

12. Pulsoximeter (SA Instruments, Stony Brook, USA).

---

3   Methods

3.1   Labeling NSCs with MRI Contrast Agent

1. Plate NSCs on a 21-cm^2 tissue culture dish with medium, e.g., a number of 1.2×10^4 NSCs cells per cm^2 (see Note 3).

2. After 48 h change the medium.

3. Add SPIO nanoparticles directly to the NSC medium at a concentration of 224 µg Fe/ml (20 µl/ml Endorem).

4. After 24 h aspirate the contrast agent/medium-mix away and wash the cells with PBS.

5. Afterwards incubate the NSCs for 5 min with Accutase at 37 °C (use 0.5 ml Accutase per 21-cm^2 tissue culture dish). Cells are now detached.

6. Mix the detached cells with 1.5 ml medium and centrifuge for 3 min at 1,000 rpm ($200 \times g$).

7. Aspirate the supernatant off and dissolve the cell pellet (see Note 4).

8. Labeled NSCs are counted and used either for viability assay, transplantation, or MRI phantom.

3.2   Evaluation of Label Influence on Cell Viability (Presto Blue Test)

1. Plate NSCs on a 96-well plate—labeled (see Section 3.1) and unlabeled NSCs in triplicate: 5,000 and 10,000 cells in 100 µl medium per well (see Note 5).

2. After 48 h aspirate the medium off and wash cells with PBS (see Note 6). Subsequently, add 90 µl of fresh medium.

3. To determine the background, add 90 µl medium in a well without cells, also as triplicate.

4. As next step, add 10 µl PrestoBlue in each well and replace the plate in the incubator.

5. Measure the plate after 2 h incubation time in a plate reader (spectrophotometer, duration: 0.5 s; lamp energy: 14,000; excitation: 530 nm; emission: 600 nm) (see Note 7).

3.3 Cell Transplantation

Preparation of cells

1. For intracerebral transplantation, culture cells at least two passages in advance to allow balanced gene expression and high cell viability. On the day of surgery, (labeled) cells are detached and counted (see Section 3.1).

2. Adjust the cell amount carefully to result in a total injection volume of 2 µl (e.g. 150,000 cells/µl for the injection of 300,000 cells) in HBSS buffer (see Note 8).

3. Keep cells on ice until transplantation to ensure sustained viability (see Note 9).

Intracerebral cell transplantation into the mouse brain (Fig. 1)

4. Weigh the animal and anesthetize with 2 % isoflurane in O_2: N_2O (30:70).

5. Place animal straight into the stereotactic frame and fix with ear sticks, ensure horizontal head alignment.

6. Insert the rectal temperature probe and check body temperature regularly during surgery.

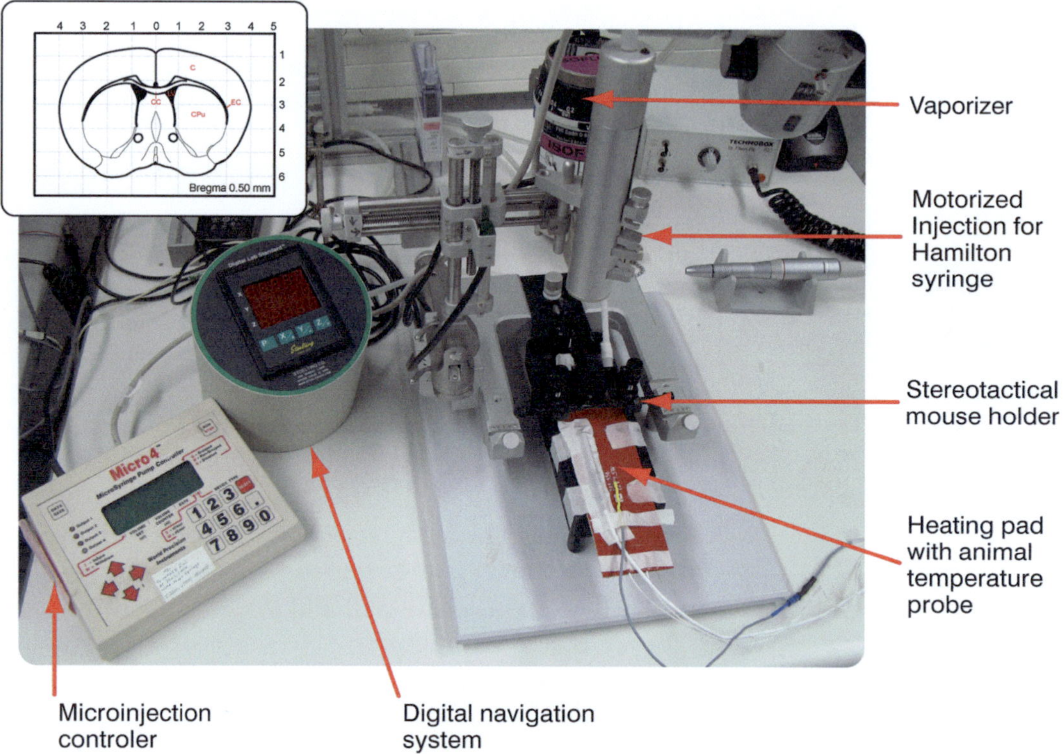

Fig. 1 Picture of the surgical equipment needed for mouse brain stereotaxis: The animal is placed on a stereotactical holder, which allows heating during surgery and the precise injection of µl-volume of cell suspension according to the coordinates of the Paxinos and Franklin's mouse brain atlas (scheme *left upper corner*)

7. Inject the analgetic drug Carprofen (4 mg/kg, s.c.) using a 2 ml syringe with a 30 G needle (see Note 10).

8. Open the scalp with a scissor until lambda and bregma appear visible, scratch remaining skin.

9. Adjust the stereotactic instrument with the Hamilton syringe to bregma (=point 0) and mark the desired coordinates with a waterproof pen (see Note 11).

10. Drill on the labels down to the dura mater slowly (!), check if the brain parenchyma becomes visible (see Note 12).

11. Draw the Hamilton syringe up (1,500 nl/min) with the (freshly) mixed cell suspension, check for first drop at the tip before injection (150 nl/min) inject cells and leave syringe in place for 5 min, remove it carefully and slowly (see Note 13).

12. Suture the wound (1.5 metric, noncolored) and let the animal wake up under controlled conditions.

3.4 Protocol for In Vitro BLI

Prior to the transplantation procedure, genetically modified cells have to be tested for consistent photon emission/cell over passages and time. Furthermore, the in vitro BLI can serve as an external control for cells intended for transplantation. The emitted photons can be recorded with the BLI camera (as it is described here), or with a plate reader for measuring luminescence (note room temperature condition and measurement in relative units and sensitivity differences compared to a CCD).

In vitro BLI protocol

1. Prepare D-luciferin stock solution (10 mM) (see Section 2.3).

2. Let the CCD camera cool down to −20 °C, start Photon acquisition software.

3. Plate cells in black 96-well plates, e.g., in a dilution series ranging from 1×10^3 to 1×10^5 cells in 90 µl medium.

4. Add 10 µl D-luciferin stock solution (1 mM final concentration) and start imaging with acquisition time of 1 min.

3.5 Protocol for In Vivo BLI (Fig. 2)

BLI of transplanted cells described here, allows repetitive measurements of the same animal for a long period of time (months) without affecting animal physiology. Care has to be taken for the intraperitoneal (i.p.) substrate injection which is a potential source of infection and inflammation. Dependent on the type of experimental question, a long-term kinetic curve for 30–60 min or a short acquisition for 1 min at the peak emission may be acquired.

In vivo BLI protocol

1. Prepare D-luciferin stock solution (20 mg/ml) (see Section 2.3).

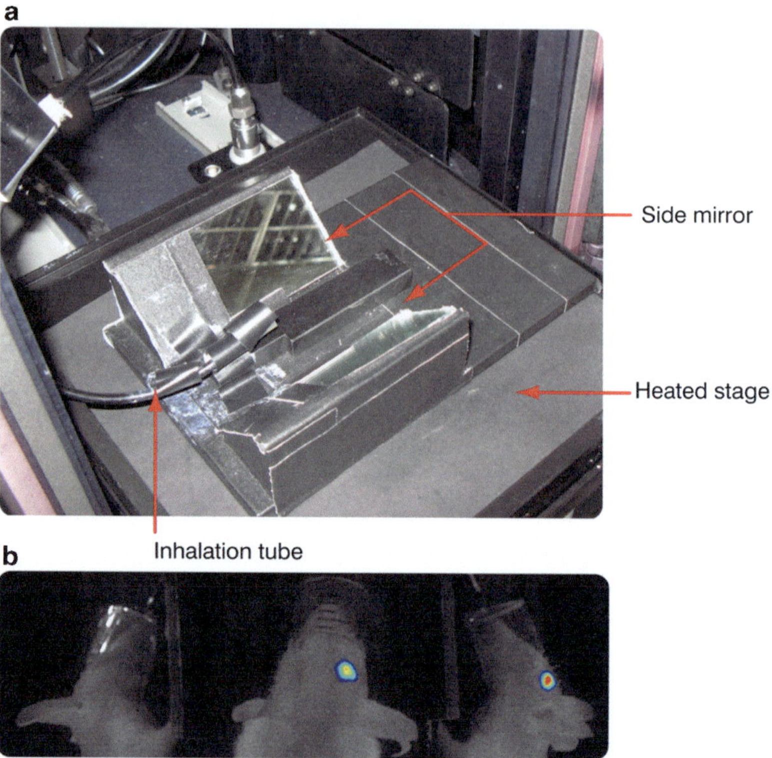

Fig. 2 Bioluminescence setup for in vivo imaging. (**a**) For mouse brain BLI, a custom-made mirror system was used, which allows consistent animal positioning and a side-view on the brain. (**b**) Representative image of a mouse receiving a Luc2+ NSC graft in the right striatum acquired with the mirror system

2. Let the CCD camera cool down to −20 °C, start Photon acquisition software.

3. Place the home-built mirror system into the Biospace so that the anesthesia mask lies left, couple the anesthesia connection to the according plug-in, orient the mirror system at the white labels on the right site.

4. Set oxygen flow to 1.5 l/min and isoflurane to 2 %.

5. Prepare the animal and check weight, behavior, and breathing.

6. Thaw luciferin stock (20 mg/ml) and inject 150 mg/kg D-luciferin i.p. into the anesthetized animal (see Note 14).

7. After positioning the animal in the Biospace optical imager, start BLI measurements by acquiring data during 30 s–1 min sessions (maximum emission is reached approx. 15 min after substrate injection).

3.6 Protocol for In Vitro MRI

Preparation of the Phantom

1. In a first step, fill 50 μl 1 % liquid agarose in a 200 μl tube. Wait until the agarose is solid (see Note 15).

2. In a second step, count the labeled cells (see Section 3.1) and dissolve the intended cell amount in 15 µl PBS, and subsequently mix with 15 µl of 1 % agarose type VII which is cooled down to room temperature (see Note 16). Fill the cell/agarose mix in the 200 µl tube on top of the solid agarose. Wait until the cell/agarose mix is also solid.

3. In a third step, add 100 µl 1 % agarose in the 200 µl tube on top of the cell/agarose mix. Wait until the agarose is solid.

4. The tubes are placed in an in-house designed tube holder. The inner part of the phantom can be filled with 1 % agarose. Hence, the cells are surrounded by agarose (Fig. 3) (see Note 17).

Prepare the MRI equipment

5. Insert the 9 cm diameter resonator for transmission.

6. Connect the rat holder.

7. Place the Phantom in the rat holder and attach the rat quadrature surface coil.

8. Insert the rat holder in the Biospec 11.7 T animal scanner system.

In vitro MRI protocol

9. Conduct a TriPilot to check the correct positioning of the phantom in the isocenter of the magnet. If necessary reposition the holder.

10. After tuning and matching of the coil, run a single-shot gradient echo-echo planar imaging (GE-EPI)-Pilot with only the basic shim to check the image quality and the degree of distortions.

11. Shim with local shim and MapShim until image quality is satisfying.

12. The imaging protocol consists of:
    - A T2*-weighted 2D multigradient echo (MGE) sequence: number of slices = 8, slice thickness = 0.5 mm, FOV = 32 × 24 mm, matrix = 128 × 96, TE/TR 3.2 ms/1,000 ms, 12 echoes, FA = 30°, TA = 4:48 min.
    - A T2-weighted 2D multi-slice multi spin echo (MSME) sequence: geometry and resolution identical to MGE, TE/TR = 10.25 ms/3,000 ms, 16 echoes, TA = 19:12 min.
    - A RARE variable time repetition (RAREVTR): geometry and resolution identical to MGE, TE/TR =13.3 ms/ 3,000 ms, 1 echo, TA = 30:23 min.

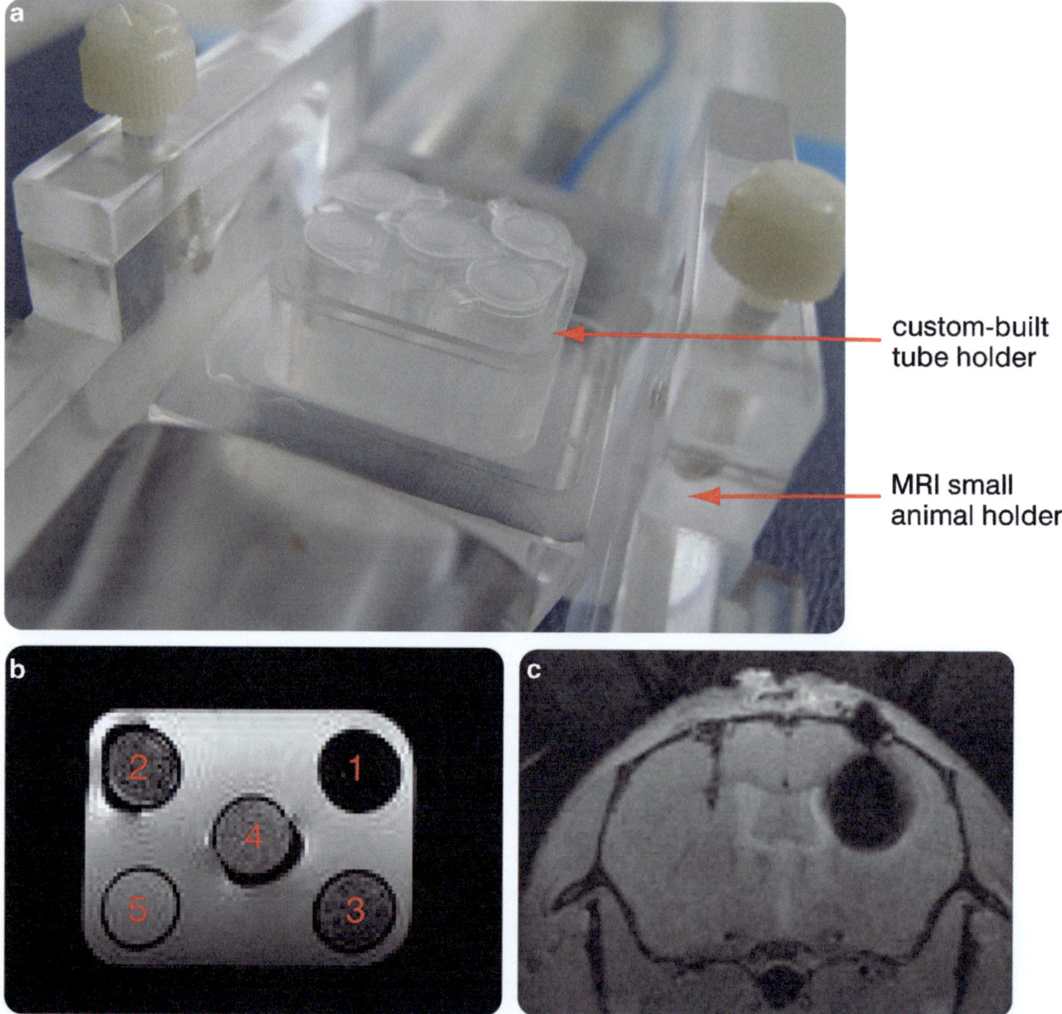

Fig. 3 Verification of SPIO-labeled NSCs. (**a**) Image of the in-house designed tube holder. (**b**) MR image of a phantom of NSCs labeled with different concentration of Endorem. The phantom contains five tubes, in each tube 5×10^4 labeled cells (*1–4*) and unlabeled cells (*5*) are mixed with 1 % agarose. Labeled NSCs with 224 µg (*1*), 112 µg (*2*), 56 µg (*3*), and 28 µg (*4*) Endorem showed hypointensity in the MR image proportional to the SPIO nanoparticle amount. Cells without contrast agent (*5*) showed no hypointensity in the MR image. (**c**) 3D MGE image of an animal grafted with SPIO-labeled NSCs and scanned directly after implantation. On the *left hemisphere* 3×10^5 cells are implanted without prior labeling while on the *right hemisphere* 3×10^5 labeled cells are implanted

3.7 Protocol for In Vivo MRI

Prepare the MRI equipment

1. Insert the 9 cm diameter resonator for transmission.

2. Connect the mouse holder and connect gas tube, water tubes, temperature cable, and respiration tube.

3. Connect the heating blanket to the holder and check that the following items are in place:

- Temperature probe.
- Breathing pad.
- Tooth bar and ear bars.
- Bepanthen® cream for the eyes.
- Pulse oximeter.

4. Turn on the temperature control unit. For mice it should regulate for 37 °C with a maximum water temperature of 43 °C to avoid burning.

Preparation of the animal

5. Weigh the animal.

6. For anesthesia use 2 % isoflurane in O_2:N_2O (30:70) (see Note 18).

7. Place the animal in the anesthesia knock-out-box and flood the induction box with isoflurane. During preparation, isoflurane can be reduced to 1.5 %.

8. Fix the animal under isoflurane in the animal holder and add eye cream.

9. Place the agar cap into the coil and the coil onto the head.

10. Place the animal holder in the Biospec 11.7 T animal scanner system.

In vivo MRI protocol

11. Conduct a TriPilot and reposition the holder if necessary.

12. After tuning and matching of the coil, run a single-shot gradient echo-echo planar imaging (GE-EPI)-Pilot with only the basic shim to check the image quality and the degree of distortions (see Note 19).

13. Shim with local shim and MapShim until image quality is satisfying.

14. When the image quality is satisfying and the animal is breathing stable at 120–130 breaths per minute, the scan can be conducted.

15. The imaging protocol consists of:

- A T2*-weighted 3D MGE sequence: 0.1 mm isotropic resolution, FOV = 16 × 10 × 14 mm, matrix = 160 × 100 × 140, TE/TR 2.9 ms/28 ms, 6 echoes, FA = 10°, TA = 19:36 min (Fig. 3).

- A T2*-weighted 2D MGE sequence: 20 contiguous 0.6-mm slices, FOV = 19.2 × 19.2 mm, matrix = 128 × 128, TE/TR 3.5 ms/4,000 ms, 32 echoes, FA = 30°, TA = 8:32 min

- A T2-weighted 2D multi-slice multi spin echo (MSME) sequence: geometry and resolution identical to MGE, TE/TR = 10.25 ms/4,000 ms, 16 echoes, TA = 8:32 min.

4 Notes

1. The two growth factors epidermal growth factor (EGF) and fibroblast growth factor-2 (FGF-2) are only stable for 1 week in the medium—prepare fresh if necessary prior to use.

2. Do not make the agar cap too thick because it increases the distance from the surface coil, which will decrease the signal to noise ratio.

3. The appropriate cell number has to be defined for each cell line.

4. Take care of the appropriate suspension for the cell pellet: for viability assay use medium, for transplantation HBSS buffer and for the phantom PBS.

5. Although a black 96-well plate provides light shielding of neighboured wells, it is preferable to leave at least one-well space between different wells.

6. The time of incubation depends on the cell line. Therefore, we recommend to try different amounts of cells in triplicate. Please note that the cells should not be confluent (maximum 90 %).

7. Time of substrate incubation depends on the cell line. Incubation time can vary between 30 min and 5 h.

8. Precise volume adjustment is a prerequisite for a correct cells/volume relation for injection. After first centrifugation step, carefully remove the supernatant with a yellow pipette tip and repeat centrifugation for half the time to spin down liquid adhered to the tube walls. The dry pellet is resuspended in HBSS by adding half the volume first, and add more HBSS until final volume is reached. Despite HBSS, PBS and plain medium can be used, although we favored HBSS because of superior cell viability.

9. Keeping cells on ice is essential to sustain equal viability over the time used for transplantation (should not exceed 2 h). Before filling the Hamilton syringe, resuspend and mix the cell suspension.

10. Injection of analgetic drugs allows stress-free surgery and allows a pain-free wake up of the animal after surgery.

Independent of the type of analgesia, potential side-effects should not be neglected (e.g., carprofen inhibits COX-2 and acts anti-inflammatory).

11. The exact coordination of bregma and lambda, and the horizontal positioning of the animal head are important. Commercially available motorized stereotaxic frames allow computer-controlled positioning of all three orthogonal axes, but are expensive.

12. Drilling induces the strongest bias from the intended coordinates and should be done with great caution.

13. When removing the needle too fast, the small injection drop will eventually be sucked out again. The only way to avoid this failed injection is leaving the needle in place for several min to allow dispersion of the injection drop and a very slow needle removal.

14. The precise and consistent injection of substrate is of importance for comparing BLI signals of different animals and time points. The substrate dose has to be determined for each animal weight. Control over injection failure could potentially be monitored by using the commercially available D-luciferin Rediject (Perkin Elmer), which includes a fluorescent marker for indication of substrate distribution.

15. It is very important to avoid air bubbles, which will lead to hypointense spots in T2-weighted images potentially leading to misinterpretations.

16. The advantage of agarose type VII is that it can be cooled down to room temperature and it is still fluid. Cells should only be mixed with room temperature-cooled agarose type VII. Avoid air bubbles!

17. Avoid air bubbles!

18. The percentage of isoflurane depends on the animal weight and the equipment. Check the breathing of the animal and adjust the percentage of isoflurane accordingly.

19. Avoid solidified agar and replace the phantom by preparing new agar solution (once agar is solidified, it cannot be heated up again).

References

1. de Almeida PE, van Rappard JR, Wu JC (2011) In vivo bioluminescence for tracking cell fate and function. Am J Physiol Heart Circ Physiol 301(3):H663–H671. doi:10.1152/ajpheart.00337.2011

2. Li Z, Suzuki Y, Huang M, Cao F, Xie X, Connolly AJ, Yang PC, Wu JC (2008) Comparison of reporter gene and iron particle labeling for tracking fate of human embryonic stem cells and differentiated endothelial cells in living subjects. Stem Cells 26(4):864–873. doi:10.1634/stemcells.2007-0843

3. Politi LS (2007) MR-based imaging of neural stem cells. Neuroradiology 49(6):523–534. doi:10.1007/s00234-007-0219-z

4. Shapiro EM, Sharer K, Skrtic S, Koretsky AP (2006) In vivo detection of single cells by MRI. Magn Reson Med 55(2):242–249. doi:10.1002/mrm. 20718

5. Cormode DP, Jarzyna PA, Mulder WJ, Fayad ZA (2010) Modified natural nanoparticles as contrast agents for medical imaging. Adv Drug Deliv Rev 62(3):329–338. doi:10.1016/j. addr.2009.11.005

6. Himmelreich U, Dresselaers T (2009) Cell labeling and tracking for experimental models using magnetic resonance imaging. Methods 48(2):112–124. doi:10.1016/j.ymeth.2009. 03.020

7. Hoehn M, Wiedermann D, Justicia C, Ramos-Cabrer P, Kruttwig K, Farr T, Himmelreich U (2007) Cell tracking using magnetic resonance imaging. J Physiol 584(Pt 1):25–30. doi:10.1113/jphysiol.2007.139451

8. Modo M, Hoehn M, Bulte JW (2005) Cellular MR imaging. Mol Imaging 4(3):143–164

9. Nieman BJ, Shyu JY, Rodriguez JJ, Garcia AD, Joyner AL, Turnbull DH (2010) In vivo MRI of neural cell migration dynamics in the mouse brain. Neuroimage 50(2):456–464. doi: 10.1016/j.neuroimage.2009.12.107

10. Saleh A, Wiedermann D, Schroeter M, Jonkmanns C, Jander S, Hoehn M (2004) Central nervous system inflammatory response after cerebral infarction as detected by magnetic resonance imaging. NMR Biomed 17 (4):163–169. doi:10.1002/nbm.881

11. Shapiro EM, Gonzalez-Perez O, Manuel Garcia-Verdugo J, Alvarez-Buylla A, Koretsky AP (2006) Magnetic resonance imaging of the migration of neuronal precursors generated in the adult rodent brain. Neuroimage 32 (3):1150–1157. doi:10.1016/j.neuro-image.2006.04.219

12. Sumner JP, Shapiro EM, Maric D, Conroy R, Koretsky AP (2009) In vivo labeling of adult neural progenitors for MRI with micron sized particles of iron oxide: quantification of labeled cell phenotype. Neuroimage 44(3):671–678. doi:10.1016/j.neuroimage.2008.07.050

13. Vande Velde G, Rangarajan JR, Toelen J, Dresselaers T, Ibrahimi A, Krylychkina O, Vreys R, Van der Linden A, Maes F, Debyser Z, Himmelreich U, Baekelandt V (2011) Evaluation of the specificity and sensitivity of ferritin as an MRI reporter gene in the mouse brain using lentiviral and adeno-associated viral vectors. Gene Ther 18(6):594–605. doi:10.1038/gt.2011.2

14. Vreys R, Vande Velde G, Krylychkina O, Vellema M, Verhoye M, Timmermans JP, Baekelandt V, Van der Linden A (2010) MRI visualization of endogenous neural progenitor cell migration along the RMS in the adult mouse brain: validation of various MPIO labeling strategies. Neuroimage 49 (3):2094–2103. doi:10.1016/j.neuroimage. 2009.10.034

15. Sutton EJ, Henning TD, Pichler BJ, Bremer C, Daldrup-Link HE (2008) Cell tracking with optical imaging. Eur Radiol 18 (10):2021–2032. doi:10.1007/s00330-008-0984-z

16. Zhao H, Doyle TC, Coquoz O, Kalish F, Rice BW, Contag CH (2005) Emission spectra of bioluminescent reporters and interaction with mammalian tissue determine the sensitivity of detection in vivo. J Biomed Opt 10(4):41210. doi:10.1117/1.2032388

17. Kim DE, Schellingerhout D, Ishii K, Shah K, Weissleder R (2004) Imaging of stem cell recruitment to ischemic infarcts in a murine model. Stroke 35(4):952–957. doi:10.1161/ 01.STR.0000120308.21946.5D

18. Kruttwig K, Brueggemann C, Kaijzel E, Vorhagen S, Hilger T, Lowik C, Hoehn M (2010) Development of a three-dimensional in vitro model for longitudinal observation of cell behavior: monitoring by magnetic resonance imaging and optical imaging. Mol Imaging Biol 12(4):367–376. doi:10.1007/s11307-009-0289-x

19. Couillard-Despres S, Vreys R, Aigner L, Van der Linden A (2011) In vivo monitoring of adult neurogenesis in health and disease. Front Neurosci 5:67. doi:10.3389/ fnins.2011.00067

Methods Molecular Biology (2013) 1052: 167–176
DOI 10.1007/7651_2013_16
© Springer Science+Business Media New York 2013
Published online: 7 June 2013

Magnetic Resonance Imaging and Tracking of Stem Cells

Hossein Nejadnik, Rostislav Castillo, and Heike E. Daldrup-Link

Abstract

To date, several stem cell labeling protocols have been developed, contributing to a fast growing and promising field of stem cell imaging by MRI (magnetic resonance imaging). Most of these methods utilize iron oxide nanoparticles (MION, SPIO, USPIO, VSIOP) for cell labeling, which provide negative (dark) signal effects on T2-weighted MR images. The following protocol describes stem cell labeling techniques with commercially available gadolinium chelates, which provide positive contrast on T1-weighted MR images, which can be advantageous for specific applications.

Keywords: Magnetic resonance imaging (MRI), Stem cells, Cell labeling, Cell tracking, Cellular imaging, ProHance®, Gadoteridol, Gadolinium chelate

1 Introduction

Stem cells represent a unique source for morphological and functional restoration of tissue defects. Preclinical and clinical trials are currently pursued to investigate the potential of various types of stem cells to repair a wide variety of pathological conditions. Autologous adipose-derived stem cells (ADSCs) and bone marrow-derived stem cells (BMSCs) provide several practical benefits including multipotent differentiation capacity, long-term self-renewal, and low immunogenicity (1). However, one of the main limitations for long-term success of stem cell transplants is our inability to recognize the fate of the transplanted cells in a timely manner. To date, a large proportion of transplanted stem cells undergo apoptosis and/or get lost from the transplantation site (2, 3) (Fig. 1). An imaging method that could visualize and track stem cells directly, noninvasively, and repeatedly in vivo could enhance our ability to develop more successful cell transplantation techniques (4, 5). MR imaging is currently the only noninvasive, repeatable diagnostic method, which can provide sub-millimeter anatomical resolution, high soft tissue contrast, and functional information in vivo without radiation exposure (6, 7). Transplanted

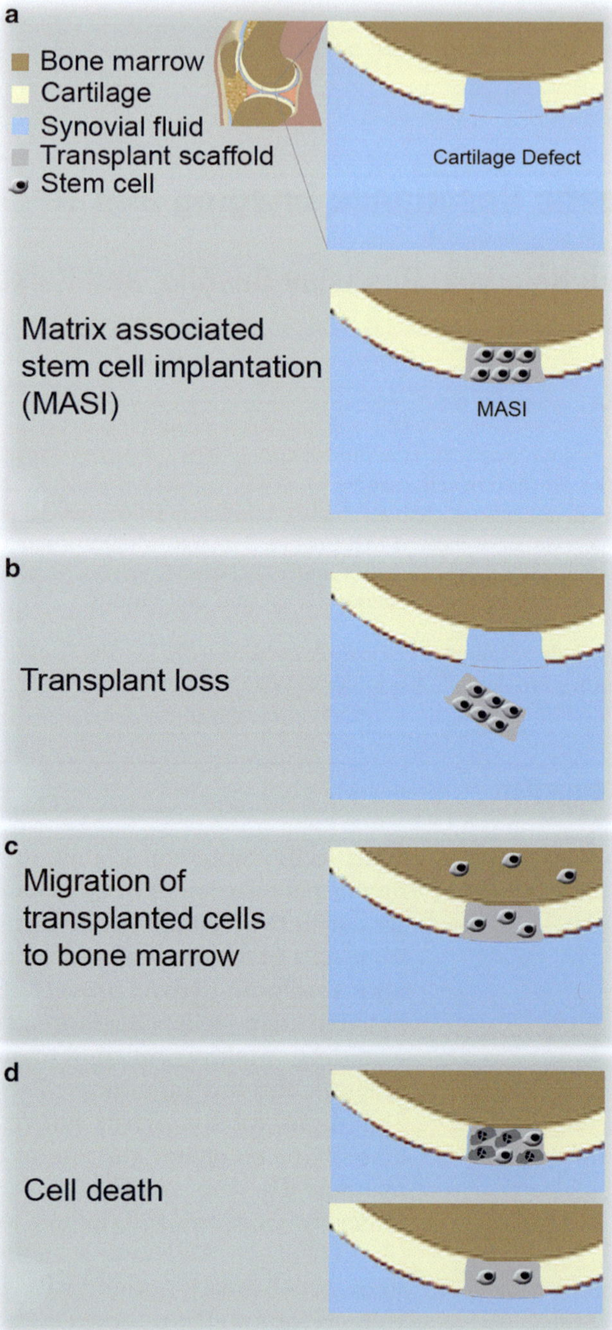

Fig. 1 Example of stem cell transplants in a cartilage defect (**a**) and potential complications, (**b**) transplant loss, e.g., due to mechanical factors, (**c**) migration of the implanted cells into bone marrow, and (**d**) cell apoptosis

stem cells can usually not be differentiated from their target organ based on intrinsic signal characteristic. They need to be labeled with contrast agents in order to track them in vivo. Gadolinium-based MR contrast agents have promising properties for MRI cell tracking, including immediate availability for clinical translations and positive (bright) contrast on T1-weighted MR images (8–12).

This chapter will provide a detailed protocol for effective stem cell labeling with standard small molecular gadolinium chelates and detection of the labeled cells with low field (1-T) and high field (7-T) MR scanners. While we used the nonionic gadolinium-based MRI contrast agent Gadoteridol (ProHance®) for the descriptions below, the same protocol can be applied for cell labeling with Gd-DTPA (Magnevist®), Gadoteric Acid (Dotarem®), Gadodiamine (Omniscan®), or other similar compounds (12–16).

2 Materials

2.1 Equipments

1. Centrifuge
2. Microscope
3. Cell culture incubator (37 °C, 5 % CO_2, 90 % humidity)
4. 20/200/1000 μL pipetter and relevant tips
5. Cell counter and slides or hemocytometer

2.2 Cell Culture

1. Adipose-derived stem cells (ADSCs) or BMSCs
2. Sterile conical tube
3. Tissue culture flask
4. Dulbecco's Modified Eagle Medium (DMEM) high glucose (Invitrogen, Carlsbad, CA, catalogue number: 11965) (see Note 1).
5. Phosphate buffered saline (PBS) 10 mM phosphate, 0.9 % NaCl, pH = 7.4 (Invitrogen, Carlsbad, CA, catalogue number: 14190) (see Note 1).
6. Trypsin 0.05 %/ethylenediaminetetraacetic acid (EDTA) 0.53 mM solution (see Note 1). Long-term storage of trypsin should be at −80 °C (Mediatech, Manassas, VA, catalogue number: 25200)
7. Fetal bovine serum (FBS, Invitrogen, Carlsbad, CA, catalogue number: 26140) (see Note 1).
8. Penicillin/Streptomycin (Invitrogen, Carlsbad, CA, catalogue number: 15140) (see Note 1).
9. Trypan blue dye (Invitrogen, Carlsbad, CA, catalogue number: 15250)

10. Collagen Type I solution (Sigma, St. Louis, MO, catalogue number: C2674)

11. Cell strainer (70 μm, BD biosciences, Bedford, MA, catalogue number: 352350)

2.3 Cell Labeling

1. ProHance® (Gadoteridol, Gd-HP-DO3A, 0.5 M stock solution) (Bracco Diagnostics, catalogue number: 0270-1111-03)

2. Lipofectin® reagent (Invitrogen, Carlsbad, CA, catalogue number: 18292-011)

2.4 Gadolinium Quantification

1. Hydrochloric acid (HCl), metal grade (Fisher Scientific, Fair Lawn, NJ, catalogue number: A508-500)

2. Filter 0.2 μm pore size (Thermo Scientific Nalgene® syringe filter, catalogue number: 190-2520)

2.5 Magnetic Resonance Imaging

1. NMR tubes 3 mm (Bruker, catalogue number: Z117723) or PCR tubes (Thermowell™ Tube, Corning Incorporation, NY, catalogue number: 6571)

2. NMR or PCR tube holder (homemade)

3. MR scanner (we used ASPECT 1 T MR scanner and Varian 7 T MR scanner with GE interface)

---

3   Methods

Labeling stem cells with MR contrast agents enables us to visualize the cells in vivo, and ensure that they have been delivered to the target site. Possible applications include evaluations of different numbers of transplanted cells, different time points of transplantations and effects of scaffolds and growth factors on tissue regeneration outcomes. In addition, in vivo cell tracking techniques enable us to detect complications of the engraftment process, such as stem cell loss, apoptosis, or rejection (Fig. 1).

The protocols described here have been optimized for labeling of adipose derive stem cells (ADSCs) and bone marrow-derived mesenchymal stem cells (MSCs) with small molecular gadolinium chelates, and related detection of the labeled stem cells with MR imaging. Labeling of other cell types may require modifications and adjustments of this protocol.

This protocol pertains to (1) stem cell isolation and culturing (2) labeling of stem cells with Gadoteridol, (3) preparation for imaging, (4) Gadolinium quantification, and (5) MR imaging of the labeled cells and post imaging analysis:

3.1 Cell Culturing

1. Isolation and expansion of adipose-derived stem cells (ADSCs).

   (a) After harvesting the adipose tissue from the rat, keep the tissue in cold PBS.

(b) Move the adipose tissue from the tubes to a sterile Petri dish.

(c) Add 5–10 mL of PBS to prevent dehydration.

(d) Cut the tissue into small pieces (<2–3 mm^3) using a sterile scalpel.

(e) To digest the tissue, transfer the homogenized adipose tissue into a conical tube, add 0.075 % Collagen Type I solution.

(f) Incubate the tube for 60 min at 37 °C with an agitation every 10–20 min.

(g) After 60–90 min of incubation, the adipose tissue will be digested and disappear.

(h) After digestion, centrifuge the tube(s) at room temperature, for 10 min at 500 × g.

(i) Gently, remove the saline supernatant and fat layer from the tube (Be careful not to remove the cell pellet).

(j) Add 10 mL PBS, resuspend the pellet, and filter the cell suspension through a 70 μm cell strainer.

(k) Centrifuge at 500 × g for 5 min at room temperature and throw away the supernatant.

(l) Resuspend the cell pellet in 10 mL of the complete cell culture medium (high glucose DMEM supplemented with 10 % FBS and 100 IU Penicillin and 100 μg/mL Streptomycin)

(m) Seed the cells in 25 cm^2 cell culture flask.

(n) After 24 h, change the cell culture medium.

2. Isolation and expansion of Human mesenchymal stem cells (hMSCs).

(a) Human bone marrow-derived mesenchymal stem cells (hMSCs) can be harvested from aspirated bone marrow (from iliac crest and/or sternum) of healthy donors or can be bought from Lonza (catalogue number PT-2501).

3. Bone marrow aspirates are diluted in DMEM and centrifuged two times for 10 min at 500 × g.

4. Cell pellet resuspended in complete medium containing high glucose DMEM supplemented with 10 % FBS and 100 IU Penicillin and 100 μg/mL Streptomycin, are plated in a 75 cm^2 cell culture flask.

5. After 48 h, to remove the nonadherent cells, wash the cells with prewarmed PBS and change the PBS with prewarmed fresh medium. The medium is replaced every 2–3 days as the cells grow to confluence (see Note 2).

3.2 Cell Labeling

Cells are labeled with Gadoteridol (ProHance®, see Note 3). In order to facilitate efficient transfection of the contrast agent into cells, the transfection agent Lipofectin® is used (see Notes 4 and 5).

1. Add 0/1/2/4 µL/mL of Lipofectin® at concentrations of 0/0.5/1/2 mM to four 1 mL tubes containing 500 µL of DMEM each. Allow the solution to sit 15 min for the Lipofectin® to distribute homogenously.

2. Add 0/1/2/4 µL/mL of Gadoteridol (ProHance®) at concentrations of 0/0.5/1/2 mM to four 1 mL tubes containing 500 µL of Lipofectin® plus DMEM each. Allow the solution to sit for 30 min more for the Lipofectin® to complex with ProHance®.

3. Prepare the labeling medium by adding 8 mL DMEM (final volume will be 9 mL).

4. Aspirate the stem cell medium and wash the cells with prewarmed PBS one time to remove any dead cells and the remaining cell medium.

5. Add the labeling medium to each flask containing the attached stem cells. Incubate the cells with the serum free labeling medium for 4 h.

6. After 4 h, supplement the labeling medium with 10 % FBS (1 mL) and 100 IU/mL Penicillin and 100 µg/mL Streptomycin (100 µL of 100× pen/strep) in each flask and continue labeling for 24 h (see Note 6).

3.3 Cell Detachment and Preparation for Imaging

1. Wash the cells in each flask with 5 mL of PBS for each flask three times (see Note 7).

2. Add 3 mL of Trypsin to each flask and place them in 37 °C incubator for 2–5 min (see Note 8).

3. Add 3 mL of complete medium to each flask to neutralize Trypsin activity.

4. Remove the 6 mL of medium from each tube and add them to their respective saved medium and PBS.

5. Spin the cells down at $500 \times g$ for 5 min. Aspirate the supernatant and resuspend in 1 mL of complete medium for each of the four labeling concentrations.

6. Mix 10 µL of Trypan blue and 10 µL of the detached cell suspension of each four samples. Gently mix the two solutions and place 10 µL of this solution onto a cell counter slide and count the number of cells per mL of cell suspension.

7. Take an equal number of cells (e.g., 2×106 cells) from each concentration to create a uniform cell pellet for MR imaging. Spin the cells down at $500 \times g$ for 5 min. Resuspend each of the samples in 80µL of complete medium.

8. Transfer the 80 μL from each sample into an NMR tube for MR imaging. Place the NMR tubes into 1 mL centrifuge tubes and spin them down at $500 \times g$ for 5 min.

3.4 Gadolinium Quantification

1. Take an equal number of cells (e.g., 1×106 cells) from each concentration to prepare the samples for Gadolinium (Gd) measurement by inductively coupled plasma optical emission spectrometry (ICP-OES).

2. Dissolve the cell pallets with 200 μL of 70 % HCl (metal grade) and incubate overnight at room temperature.

3. Top up the samples with DI water to 5 mL.

4. Filter the samples with a 0.2 μm pore size filter.

5. Prepare your blank (DI water) and low concentration (1 ppm) and high concentration (2 ppm) Gd standards and send the samples and standards for ICP-OES.

3.5 Cell Imaging and T1 Relaxation Time Calculation

1. Place each NMR tube into a slot in an MR tube holder filled with fluid (e.g., PBS). Put the holder in an MR coil and place the holder and coil in the center of the magnet of the MR scanner.

2. Scan the cells with a T1-weighted spin echo (T1W SE) or gradient echo (GE) sequence to visualize the gadolinium-labeled cells. Parameters for a T1-weighted spin echo (T1W SE) sequence are 500 ms repetition time (TR); 13–17 ms echo time (TE); 90° flip angle (FA); 256×256 image matrix; 0.5–1 mm slice thickness (ST): and 8 number of excitation (NEX) (Fig. 2). These parameters may have to be adjusted, depending on the applied field strength, medium, background signal, number of cells, etc.

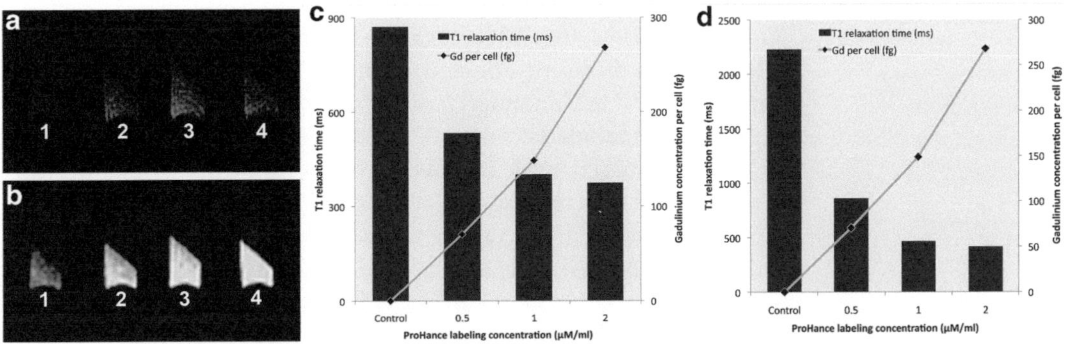

Fig. 2 Sagittal MR Imaging of pellets of 2×106 unlabeled (*left*) and labeled ADSCs (0.5, 1, and 2 mM Gadoteridol) in test tubes: (**a**) T1W SE 500/17 (TR/TE) at 7 T, (**b**) T1W SE 500/13 (TR/TE) at 1 T; panels (**c, d**) show corresponding T1 relaxation times of labeled and unlabeled ADSCs, determined based on IR sequences, as well as corresponding Gd content of the labeled cells, as measured by inductively coupled plasma (ICP)

3. To calculate T1 relaxation times of labeled cells, an inversion recovery spin echo (IR-SE) sequence with varying inversion times (TI) can be used. Examples of parameters for our typical IR-SE are repetition time (TR) 8,000 ms; echo time (TE) 8 ms; and TI 4,000, 2,000, 1,000, 500, 250, 100, 50, 25 ms; FA of 90°; image matrix 128 X 128; slice thickness (ST) 0.5–1 mm: and 1–2 number of excitation (NEX).

4. After completion of the MR scan, images should be imported to a DICOM viewer for post-processing with specific software, such as Osirix software (free download via the Internet).

5. The signal intensity for each cell pellet can be measured via an operator defined region of interest (ROI).

6. The data can be further analyzed with excel software to calculate T1 relaxation times of the labeled cells, by using the following algorithm:

$$(\ln(2) - \ln(1 - (SI_{each\ TI}/MaxSI))) \times 1,000$$

where ln is the logarithm to the base e (natural logarithm), SI is the signal intensity, TI is inversion recovery time, and MaxSI is the absolute maximum signal intensity between the inversion times.
Then, the new value for each inversion recovery is calculated.

7. Fit a linear curve to the calculated values, use the slope of the equation as R1 values and calculate the T1 relaxation times by the following formula:

$$T1 = 1/R1$$

4 Discussion and Conclusion

Figure 2 shows the T1 relaxation times of ADSCs, which were labeled with different concentrations of Gadoteridol and Lipofectin. The ICP-OES analysis of our samples showed no detectable Gd content in unlabeled control ADSCs, and an increasing Gd content in cells incubated with increasing concentrations of Gadoteridol. As expected, increasing Gadoteridol incubation concentrations yielded shorter T1 relaxation times at both 1 T and 7 T magnetic field strengths. In conclusion, we described a method for labeling stem cells with commercially available small molecular gadolinium chelates (ProHance®/Gadoteridol), which provide positive (bright) signal effects on T1-weighted MR images.

5   Notes

1. DMEM, PBS, and Trypsin should be prewarmed to 37 °C in a water bath before use.

2. The cells are lifted by incubation with 0.5 % Trypsin. The first passage after plating is usually taken as P0. Cells can be expanded up to passage 6 (P6) though the efficiency of cell labeling decreases with the number of passages.

3. ProHance® (Gadoteridol), is a gadolinium-based contrast medium that belongs to the MRI class of small molecular paramagnetic contrast agents (ATC Class V08CA). Each milliliter of ProHance® contains 279.3 mg Gadoteridol, 0.23 mg Calteridol calcium, 1.21 mg tromethamine and water. Gadoteridol is the active pharmaceutical ingredient of ProHance®, with a molecular structure of a macrocyclic, nonionic, and high stability chelate. ProHance® contains no antimicrobial preservative and has a pH of 6.5–8.0.

4. Lipofectin®, a reagent composed of the cationic lipids DOTMA and DOPE in a 1:1 mixture in membrane-filtered water, is used to transfect RNA or DNA. However, Lipofectin® transfection has also been used in labeling with contrast agent (17, 18).

5. Negatively charged MR contrast agents likely complex with positively charged lipid molecules. After fusing to the cell membrane these complexes release the MR contrast agents into the cytosol (19, 20).

6. Alternatively, the cells may be labeled for 48 h instead of 24.

7. Under a microscope, observe the cells to determine that they do not detach from the flask. If cells are detached, stop washing or use the complete prewarmed medium instead of PBS.

8. Before proceeding to the next step, check the cells under the microscope to determine if they are detached from the flask. If they are still attached, place the flasks in an incubator for another 2 min to allow the Trypsin to cleave the attachments. Repeat until more than 70–80 % of the cells are detached.

References

1. Yuan L, Sakamoto N, Song G, Sato M (2012) Migration of human mesenchymal stem cells under low shear stress mediated by MAPK signaling. Stem Cells Dev 21(13):2520–2530

2. Berman SM, Walczak P, Bulte JW (2011) MRI of transplanted neural stem cells. Methods Mol Biol 711:435–449

3. Biancone L, Crich SG, Cantaluppi V et al (2007) Magnetic resonance imaging of gadolinium-labeled pancreatic islets for experimental transplantation. NMR Biomed 20:40–48

4. Modo M, Meade TJ, Mitry RR (2009) Liver cell labelling with MRI contrast agents. Methods Mol Biol 481:207–219

5. Sykova E, Jendelova P, Herynek V (2011) Magnetic resonance imaging of stem cell migration. Methods Mol Biol 750:79–90

6. Kedziorek DA, Kraitchman DL (2010) Superparamagnetic iron oxide labeling of stem cells for MRI tracking and delivery in cardiovascular disease. Methods Mol Biol 660:171–183

7. Hedlund A, Ahren M, Gustafsson H et al (2011) Gd_2O_3 nanoparticles in hematopoietic cells for MRI contrast enhancement. Int J Nanomedicine 6:3233–3240

8. Kraitchman DL, Kedziorek DA, Bulte JW (2011) MR imaging of transplanted stem cells in myocardial infarction. Methods Mol Biol 680:141–152

9. Major JL, Meade TJ (2009) Bioresponsive, cell-penetrating, and multimeric MR contrast agents. Acc Chem Res 42:893–903

10. Caravan P (2006) Strategies for increasing the sensitivity of gadolinium based MRI contrast agents. Chem Soc Rev 35:512–523

11. Villaraza AJ, Bumb A, Brechbiel MW (2010) Macromolecules, dendrimers, and nanomaterials in magnetic resonance imaging: the interplay between size, function, and pharmacokinetics. Chem Rev 110:2921–2959

12. Tweedle MF (1997) The ProHance story: the making of a novel MRI contrast agent. Eur Radiol 7(5):225–230

13. Henning TD, Saborowski O, Golovko D et al (2007) Cell labeling with the positive MR contrast agent Gadofluorine M. Eur Radiol 17:1226–1234

14. Henning TD, Wendland MF, Golovko D et al (2009) Relaxation effects of ferucarbotran-labeled mesenchymal stem cells at 1.5 T and 3 T: discrimination of viable from lysed cells. Magn Reson Med 62:325–332

15. Henning TD, Gawande R, Khurana A et al (2012) MR imaging of ferumoxides labeled mesenchymal stem cells in cartilage defects: in vitro and in vivo investigations. Mol Imaging 11(3):197–209

16. Nejadnik H, Henning TD, Boddington S et al (2012) Somatic differentiation and MR imaging of magnetically labeled human embryonic stem cells. Cell Transplant 21(12):2555–2567

17. Rudelius M, Daldrup-Link HE, Heinzmann U et al (2003) Highly efficient paramagnetic labelling of embryonic and neuronal stem cells. Eur J Nucl Med Mol Imaging 30:1038–1044

18. Nejadnik H, Henning TD, Thuy D et al (2012) MR imaging features of gadofluorine-labeled matrix-associated stem cell implants in cartilage defects. PLoS One 7(12):e49971

19. Daldrup-Link HE, Rudelius M, Oostendorp RA et al (2003) Targeting of hematopoietic progenitor cells with MR contrast agents. Radiology 228:760–767

20. Felgner PL, Gadek TR, Holm M et al (1987) Lipofection: a highly efficient, lipid-mediated DNA-transfection procedure. Proc Natl Acad Sci U S A 84:7413–7417

Methods Molecular Biology (2013) 1052: 177–193
DOI 10.1007/7651_2013_13
© Springer Science+Business Media New York 2013
Published online: 4 June 2013

Whole Body MRI and Fluorescent Microscopy for Detection of Stem Cells Labeled with Superparamagnetic Iron Oxide (SPIO) Nanoparticles and DiI Following Intramuscular and Systemic Delivery

Boris Odintsov, Ju Lan Chun, and Suzanne E. Berry

Abstract

Methods to monitor transplanted stem cells in vivo are of great importance for potential therapeutic applications. Of particular interest are methods allowing noninvasive detection of stem cells throughout the body. Magnetic resonance imaging (MRI) is a tool that would allow detection of cells in nearly any tissue in the body and is already commonly used in the clinic. MRI tracking of stem cells is therefore feasible and likely to be easily adapted to patients receiving donor cells. Patients with Duchenne muscular dystrophy are good candidates for stem cell therapy, given the naturally regenerative nature of skeletal muscle, which repairs damage by employing endogenous stem cells from the muscle interstitium to regenerate muscle fibers throughout adulthood. We describe methods for labeling stem cells with superparamagnetic iron oxide nanoparticles (SPIO) to enhance MRI contrast, injecting them locally into skeletal and cardiac muscle, or systemically in mouse models for Duchenne muscular dystrophy, and tracking them in muscle tissue of live mice following injection. We focus on the use of whole body MRI to detect stem cells, as this is necessary for conditions such as muscular dystrophy, in which affected tissues are present throughout the body and systemic delivery of stem cells may be necessary. Emphasis is placed on the development of an MRI coil that is field of view (FOV) adjustable and can be used for both whole body imaging to determine stem cell localization as well as subsequent focusing on smaller, local regions where stem cells are present to obtain high-resolution images. We discuss the coil design and its significance for stem cell tracking. We also describe methods for labeling stem cells with a fluorescent dye and for tracking them in postmortem tissue specimens with fluorescent microscopy to correlate, compare, and contrast with results of whole body MRI in preclinical studies.

Keywords: Stem cells, Magnetic resonance imaging, Duchenne muscular dystrophy, Skeletal muscle, Cardiac muscle, Feridex, FeREX, SPIO, DiI, Fluorescence microscopy, Macrophage

1   Introduction

Stem cell implantation is a promising strategy for the treatment of many degenerative conditions. Noninvasive imaging of transplanted stem cells will be necessary in clinical trials, and is also useful for tracking stem cells in preclinical studies with animal models for human degenerative diseases. Noninvasive imaging can be used for establishing cell localization and migration periodically in the same animal following injection, and is also a rapid method for optimizing the timing of administration of stem cells based on their localization and homing to damaged tissue. MRI is well suited as an imaging modality for noninvasive cell tracking because it can be used for tissue characterization and has excellent image quality, high spatial resolution, flexible image contrast, and penetration depth, and also is advantageous because it does not require ionizing radiation (1). An increase of magnetic field strength in MRI experiments enhances NMR signal, in turn allowing for use of smaller voxels for analysis and enhancing both sensitivity and spatial resolution for detection of labeled stem cells. MRI provides regional contrast based on differences in proton density, flow, and biochemical structure, which alter regional signal intensity within acquired images. The regional contrast is detected as either negative (dark) or positive (bright) signals. In order to improve contrast further, contrast agents are used. Labeling stem cells with contrast agents such as magnetic nanoparticles is therefore one mechanism for cell tracking by MRI. superparamagnetic iron oxide (SPIO) particles are MRI contrast agents (2) that enhance contrast by creating a large dipolar magnetic field gradient that is experienced by protons in close proximity to the particle. The local magnetic gradient causes proton NMR signal dephasing related to local field inhomogeneity induced in water molecules near the nanoparticle or nanoparticle-containing stem cell (susceptibility effect). This effect is seen as negative contrast on T_2-weighted and $T_2{}^*$-weighted images due to shortening of T_2 and $T_2{}^*$ proton transverse relaxation times. Transverse relaxation in the presence of magnetic nanoparticles is a combination of "true" T_2 relaxation and relaxation $T_2{}^*$, caused by magnetic field inhomogeneities. Bowen et al. (3) demonstrated that $T_2{}^*$ sensitivity to iron-oxide-loaded cells is 70 times greater than for T_2-weighted images. At the same time T_2 contrast can provide better anatomical image quality than $T_2{}^*$ (4). Only gradient echo (no 180° rephrasing) pulse sequences can be used for $T_2{}^*$ contrast. SPIO particles also affect longitudinal (spin–lattice) T_1 relaxation of nearby water molecules, but this effect can be seen only at very high SPIO concentrations (1). Because each iron oxide nanoparticle produces a relatively large susceptibility $T_2/T_2{}^*$ effect, this is a sensitive way for determining the location of nanoparticle-labeled stem cells. We have developed a protocol for labeling myogenic stem cells with SPIO and fluorescent dyes, injecting them into muscle tissue, and tracking them with MRI in living animals over time. We have developed animal holders and an MRI coil (5) to facilitate detection of SPIO-labeled stem cells in vivo in mice, resulting in enhanced images of both the whole body and specific local regions of stem cell localization, as well as decreased time required for imaging, which is important for imaging labeled cells in live subjects. We detail our protocol with emphasis on these parameters and also provide information for labeling cells with a fluorescent dye for analysis of postmortem tissue to verify cell localization initially detected with MRI in live animals. Detection of the stem cells in postmortem tissue, using the fluorescent dye, is also important for examining donor cell fate in host tissues. Methods for injection of the stem cells into skeletal and cardiac muscle, as well as systemic injection into the chamber of the left ventricle are also provided. Using this protocol and coil, it is possible to detect as few as 100 SPIO-labeled stem cells following injection (4).

2 Materials

2.1 Iron Oxide Nanoparticles: See Notes Regarding Sources

Feridex, also known as Endorem, is available in Japan (Eiken Chemical Co., Ltd.), the EU and Brazil (Guerbet S.A.), but is no longer available in the United States. However, FeREX, a similar preparation of iron-oxide nanoparticles to Feridex is available through BioPAL (Worcester, MA, USA). This preparation is identical to Feridex, except that it does not contain citrate.

2.2 4 % Potassium Ferrocyanide Solution for Prussian Blue Staining

2 g $K_4Fe(CN)_6 \cdot 3H_2O$ in 50 mL of dH_2O

2.3 4 % Hydrochloric Acid Solution for Prussian Blue Staining

5.26 mL of 38 % HCl in 44.74 mL of dH_2O

2.4 2 % Aqueous Potassium Ferrocyanide–Hydrochloric Acid Solution for Prussian Blue Staining

1:1 ratio of 4 % Potassium Ferrocyanide and 4 % Hydrochloric Acid solutions

2.5 FOV- Adjustable RF Coil

MRI coils for use in rodents can be purchased from M2M Imaging (Cleveland, OH; http://www.m2mimaging.com/index.html). Alternatively, the tunable transmit/receive Radiofrequency (RF) coil (5) (see Note 2) used to generate images of both the whole mouse body and also detailed images of individual regions of the body that are shown in this chapter may be obtained through the Biomedical Imaging Center at the University of Illinois (http://bic.beckman.illinois.edu/resources.html). Additional information regarding construction of radiofrequency coils for MRI of small animals can be found in a review by Doty et al. (6).

2.6 MRI Magnet and Scanner

We have used a vertical bore imaging scanner (Oxford Instruments, Abington, UK) equipped with a Unity/Inova console (Varian, Palo Alto, CA), a vertical bore 14.1 T magnet (89 mm) and a 600 MHz Varian Unity/Inova NMR spectrometer equipped with four identical receiver channels and gradient coils with maximum strength of 95 G/cm, providing in-plain imaging resolution up to 10 μm (http://bic.beckman.illinois.edu/resources.html#varian). The maximum usable magnet diameter is 30 mm. MRI magnets with a strength of 1.5, 4.7, and 7 T have also been used for tracking stem cells in vivo (7–9), although higher magnetic field strength correlates directly with higher resolution images.

2.7 Anesthesia Equipment

The EZ-Anesthesia system (Euthanex Corp., Palmer, PA) is a small, portable, MRI-compatible unit. The system delivers a precise blended mixture of Oxygen and Isoflurane into a lid unit that covers a plastic induction chamber. The lid unit includes an air filter (EZ-258, Omnicon f/air, Bickford Inc., NY) canister that releases safe, filtered air back into the room.

2.8 Animal Holder for Imaging Live Animals with MRI

Details are provided for constructing a holder that will secure rodents and interface with anesthesia equipment during MRI of live mice (Fig. 1; see Notes 3 and 4). The specific holder detailed may also be obtained through the Biomedical Imaging Center at the University of Illinois (http://bic.beckman.illinois.edu/resources.html).

3 Methods

All protocols have been used with Aorta-derived mesoangioblasts (ADM) (4, 10, 16) stem cells. However, these protocols can be adapted for use with any type of stem cell.

3.1 Labeling Stem Cells with Vybrant DiI

1. Culture stem cells on gelatin-coated dishes or flasks until the cells are at a density of 75 %.

2. Trypsinize the cells using 1 mL of trypsin/EDTA for a 75 mm flask. As cells detach from the surface, gently pipet them to break up any clusters of cells and obtain a single-cell suspension.

3. Stop the trypsin reaction by adding 5 mL of complete medium with serum. Transfer the cells to a conical tube to keep the cells in suspension and prevent them from re-attaching to the flask or dish. Mix well by gently pipeting.

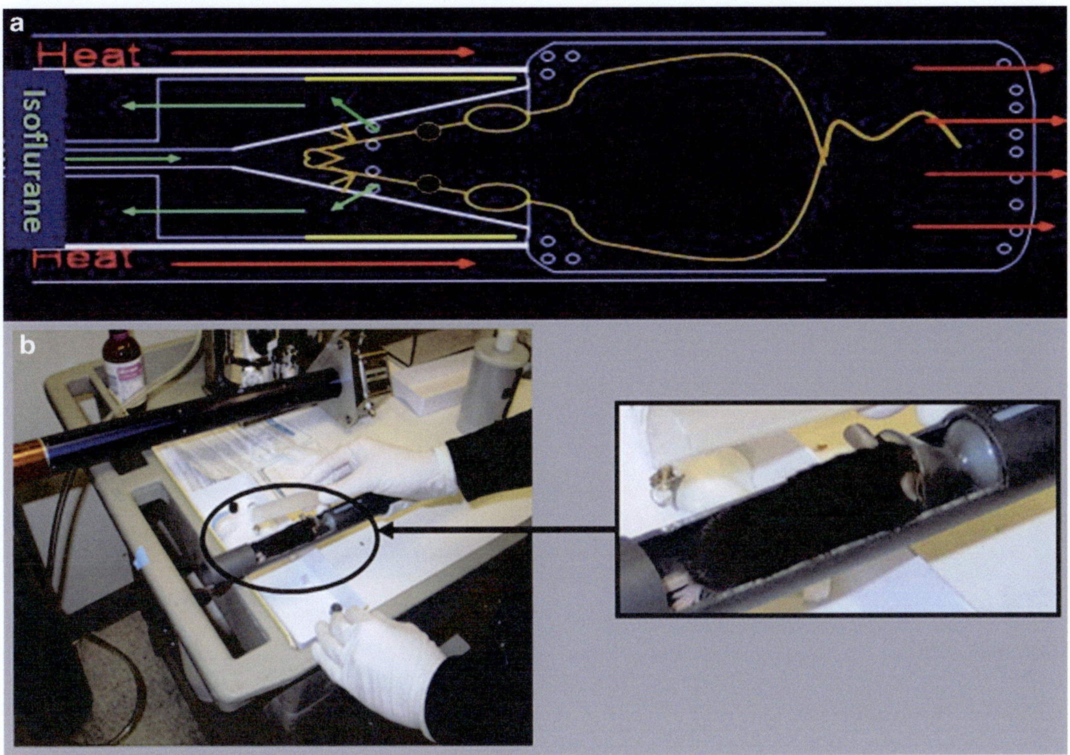

Fig. 1 An animal holder designed for whole body imaging of mice following stem cell injection. (**a**) The flow of oxygen/isoflurane (indicated by the *green arrows*) and heat (indicated by *red arrows*) pass through different channels to avoid interference. (**b**) The holder containing the mouse fits securely within the radiofrequency coil

4. Count the cells in the solution.

5. Prepare a cell suspension containing 1×10^6 cells per 1 mL of complete medium.

6. Add 5 µL of DiI (Invitrogen, Life Technologies, Grand Island NY) per mL of cell suspension, mix well by gently pipeting. DiI is light sensitive, so turn off the light in the hood when labeling cells with DiI.

7. Place cell suspension into a 37 °C incubator in the conical tube for 30 min. Gently mix cells and DiI after 15 min and place back in the incubator for the remaining 15 min. Again, be sure to avoid exposing the mixture of cells and DiI to light for prolonged periods of time.

8. Spin down cells at $3,000 \times g$, remove the supernatant, and resuspend the pellet in 10 mL of 1× Hanks balanced salt solution (HBSS) to wash the cells. Repeat this process three times to remove excess DiI and fetal bovine serum from the cell suspension (see Note 5).

9. After the final spin, resuspend the cells to the desired concentration for injection. (We have injected 1.0×10^6 ADM per 10–15 µL of HBSS) into skeletal and cardiac muscle (4, 10).

10. It is useful to quantitate the number of cells labeled with DiI to optimize labeling conditions prior to injecting cells in vivo (see Note 6 regarding labeling of cells with DiI.). After labeling cells with DiI, plate 50,000 cells in a 60 mm dish containing 2–3 gelatin-coated glass coverslips (12 mm, cat# 12-545-82 for cell culture, Fisher Scientific, Pittsburgh, PA). Fix cells with 3.7 % formaldehyde for 10 min at room temperature. Wash three times with 1× PBS, and mount on a slide using Vectashield mounting medium with DAPI (Vector Laboratories, Burlingame, CA) and view with the fluorescent microscope (Fig. 2). Quantify the percentage of DiI + cells, using DAPI to assess the total number of cells, and colocalization of DiI and DAPI to assess the number of DiI-labeled cells. Use 3–4 random fields of view to obtain quantification of labeled cells. Ideally, >95 % of cells will be labeled with DiI. The percentage of labeled cells may be increased by prolonging the incubation time and/or concentration of DiI during labeling.

3.2 Labeling Stem Cells with SPIO

1. If using ADM, coat the dishes or flasks with 0.1 % gelatin in advance, and wash the plate once with 1× PBS after coating with gelatin, before adding medium and cells.

2. Plate cells at a density of 500,000 cells per 100 mm dish, ensuring that the cells are well-separated rather than in clusters. Allow cells to attach and spread, typically overnight for ADM, prior to adding Feridex or Fe*REX* (see Note 7).

3. Dilute the SPIO stock solution to a working concentration of 100 µg/mL in the tissue culture medium. Mix well and add to cells.

4. Culture the cells in growth medium + Feridex for 3 days. If cells become more than 80 % confluent, expand them into an additional flask by trypsinization and then add more medium + Feridex.

5. Check the labeling efficiency using Prussian blue staining, and adjust the time and dose of iron oxide nanoparticles accordingly. SPIO may not be readily taken up by some stem cells. See Note 8 regarding the use of transfection agents to enhance cell labeling with SPIO.

3.3 Prussian Blue Staining for Detection of Iron Oxide Nanoparticles in Cells and Tissues

Prussian blue staining is useful for determining the percentage of cells labeled with iron particles, as well as the intensity of the staining, to ensure optimal labeling conditions prior to injection into mice (Fig. 3).

1. Incubate cells or tissues with freshly prepared (less than 10 min prior to use) 2 % aqueous potassium ferrocyanide–hydrochloric acid solution for 15–30 min, until a blue color is evident. 200 µL of solution is sufficient for each slide. Use of an ImmEdge Pen (Vector Laboratories, Burlingame, CA) to

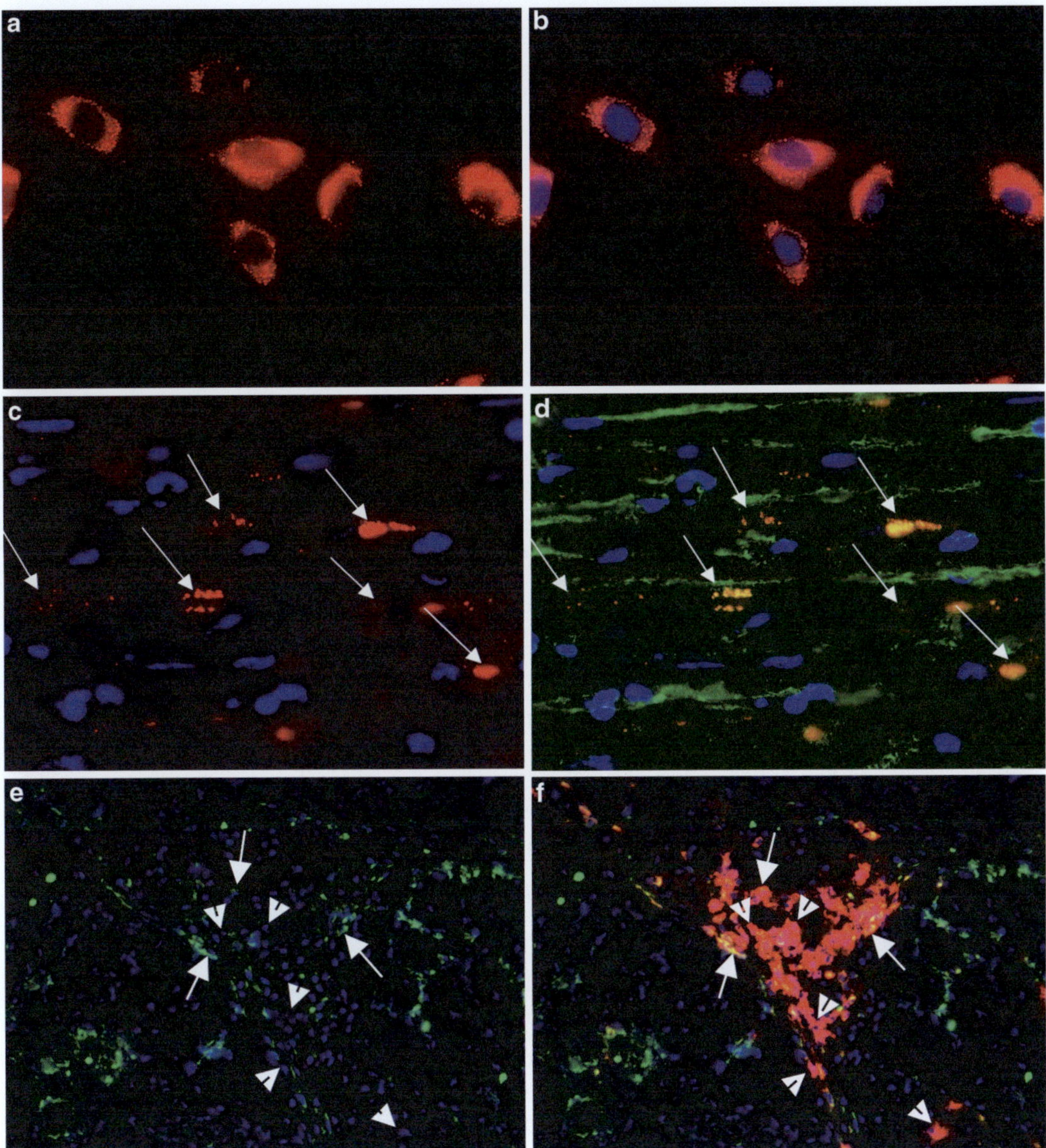

Fig. 2 Dil-labeled stem cells in vitro and in vivo. (**a, b**) Dil (red) is present in a punctate pattern in the membrane of mesoangioblasts prior to injection. (**b**) Merged image of Dil with DAPI. (**c, d**) Dil-labeled mesoangioblasts in cardiac muscle following injection into the heart wall. Cardiac tropomyosin (*green*, **d**, merged image, *arrows*) is detected in Dil-labeled cells. (**e, f**) CD68[+] macrophages (*green, arrows*) colocalize with approximately 15 % of mesoangioblasts (*red, arrowheads*) in muscle 5 weeks following injection into the gastrocnemius. (**a, b**) 60× magnification, (**c, d**) 40× magnification, (**e, f**) 20× magnification. Panels (**c–f**) reprinted from Fig. 5 of Odintsov et al., Magn Res Med 66(6): 1704–1714, 2011

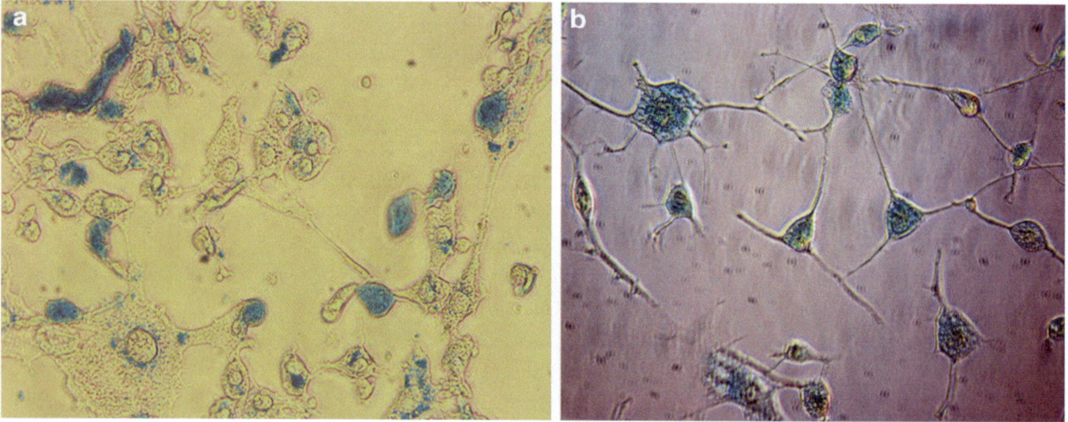

Fig. 3 Prussian blue staining for iron particles in mesoangioblasts (**a**), iron particles within undifferentiated stem cells (**b**), iron-labeled cells differentiate into oligodendrocytes when cultured in appropriate medium. Reprinted from Fig. 1 of Odintsov et al., Magn Res Med 66(6):1704–1714, 2011

generate a border around the tissue on the slide is helpful to prevent the solution from running off of the edge of the slide.

2. Wash the slides or cells in dH_2O twice.

3. Incubate with 70 % ethanol for 10 s.

4. Incubate slides in 95 % ethanol for 10 s.

5. Incubate slides in 100 % ethanol for 10 s.

6. Clear tissue with Xylene for 5 min.

7. Mount coverslips using Permount (Fisher Scientific, Pittsburgh, PA).

3.4 Preparing Cells for Injection into Mice

1. Remove complete medium, and wash cells twice with 10 mL of 1× PBS to remove serum.

2. Trypsinize labeled cells, using 0.25 % Trypsin/EDTA. Add 1× Trypsin/EDTA mixture (0.25 % Trypsin), tip the dish or flask to ensure that the trypsin/EDTA mixture has coated the surface of the container, and then place in the 37 °C incubator.

3. Remove flask or dish from the incubator in 2–3 min to determine whether cells are still adherent to the surface of the flask. If the cells remain attached return the plate to the incubator for another 2–3 min.

4. If the cells are detached, pipet them gently 5–10 times with a 5 or 10 mL pipet to break down clusters of the cells and facilitate production of a single-cell suspension. Take care not to generate bubbles.

5. Add an equivalent volume of complete medium to the cells to stop the trypsin reaction. Pipet the medium/cell solution 4–5 times to evenly distribute the cells.

6. Wash the cells to remove medium and trypsin. It is important to wash the cells well so that you do not inject serum with the cells when injecting the stem cells into animals, as this may generate an immune reaction. Centrifuge the cells, and then gently resuspend the cells in 5–10 mL of 1× PBS. Repeat centrifugation and resuspension 4 more times. Remove 20–30 µL of cell solution to determine the cell density.

7. Use a hemocytometer to determine the cell density. (we inject between 1.0×10^6 and 5×10^6 cells to detect large areas of regenerating fibers in skeletal muscle tissue)

8. Inject cells in the smallest volume possible to minimize damage to tissue. We inject cells at a density of 1×10^6 cells per 10 µL of sterile 1× HBSS solution.

9. Place cell solution for injection in a sterile Eppendorf tube on ice until injection procedure.

3.5 Injection of Cells into Skeletal Muscle

We use the gastrocnemius muscle as the skeletal muscle of choice for stem cell delivery in murine models for Duchenne muscular dystrophy (4, 10), as it is a large muscle and the degeneration and regeneration events have been well-characterized in this muscle (11, 12).

1. Anesthetize mice prior to injection (see Note 9). We have used halothane (Halocarbon, USA) mixed with mineral oil to reduce the pressure and decrease potential overdose from the anesthesia, using a mixture of 7.5 mL halothane/40 mL of mineral oil. Isoflurane may be used in place of halothane, as production of halothane has been discontinued. To prepare a nosecone with either the halothane/mineral oil mixture or isoflurane, soak a paper towel with the anesthetic, and place the paper towel at the end of a 50 mL conical tube.

2. Using a sterile insulin syringe with a 29 G needle (see Note 10 regarding the size of needles and use with different stem cells) to minimize damage to tissue, draw the cells into the syringe. Insert the syringe into the muscle and inject cells as you withdraw the syringe to distribute them more widely and minimize tissue damage rather than injecting them as a bolus in one region. For injection of the gastrocnemius, for example, the needle is inserted near the knee and then longitudinally through the muscle until the tip is near the ankle. The needle should then be withdrawn slowly and the cells administered as this occurs, to distribute them throughout the muscle (Fig. 4b–h; reproduced from Odintsov et al., 2011 (4) with permission from John Wiley and Sons).

3. For postprocedural pain management, after injection of stem cells into skeletal muscle, lidocain is administered locally.

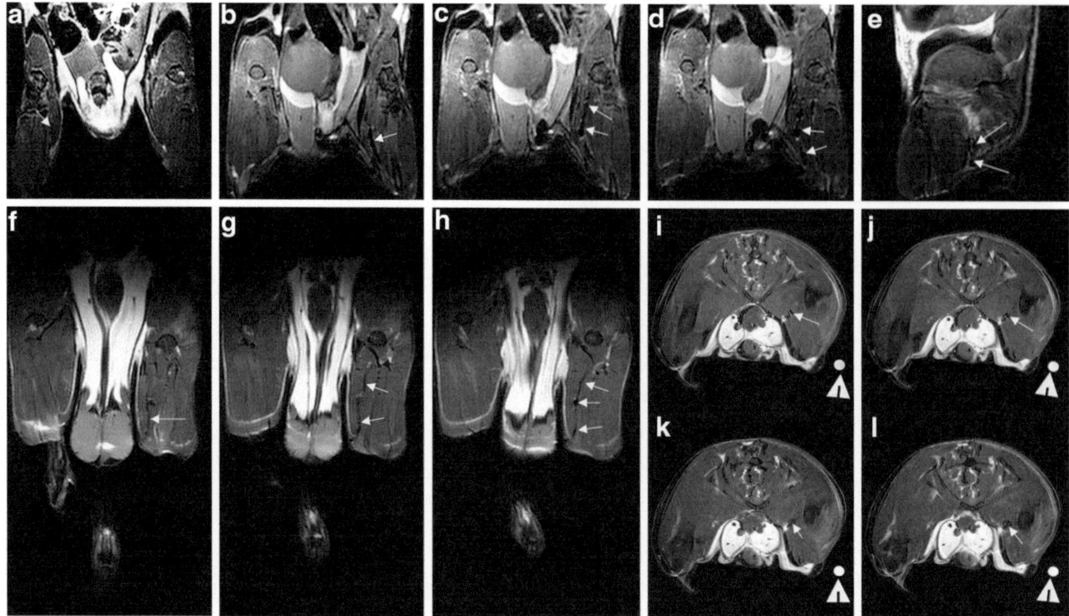

Fig. 4 MRI detection of iron-labeled mesoangioblasts injected into skeletal muscle. (**a–e**) Coronal/sagittal sections (FOV: 60 × 30 mm) (**a**), as few as 100 labeled cells are detectable (*arrow*) 24 h following injection (**b–e**), detection of five million cells (injection track indicated by *arrows*) 24 h following delivery in coronal and sagittal sections of an mdx mouse. A hypointense region is detected in the leg injected with stem cells (indicated by the *arrows*) but not in the contralateral leg injected with an equal volume of saline solution (**b–d**), serial images of consecutive sections (**f–l**), serial imaging of iron-labeled stem cells 2 weeks following injection in both coronal and sagittal sections of the hindlimbs (resolution—100 μm) (**f–h**), coronal sections (FOV: 30 × 30 mm), (**i–l**) transverse sections (FOV: 30 × 30 mm). *Arrowheads* in (**i–l**) indicate the phantom placed next to the leg injected with stem cells to distinguish it from the saline-injected muscle. MRI parameters: slice thickness—0.5 mm; repetition time (TR)—2 s; matrix size—256 × 256; TE = 25 ms. Figure and legend reprinted from Odintsov et al., Magn Reson Med 66(6):1704–1714, 2011

3.6 Injection of Stem Cells into the Chamber (Systemic Injection) or Wall of the Heart

This method of injection into the heart is performed without opening the chest cavity. It is therefore less invasive for the subject, and is necessary when performing cell injection into animal models that are weak or very ill that may not survive opening the body cavity. This is true for the $mdx/utrn^{-/-}$ double knockout mouse model for Duchenne muscular dystrophy (11, 12). However, this method of injection is also very difficult to perform with accuracy. It is recommended to use a dye such as Evans blue dye to practice the procedure on nonexperimental animals prior to injecting cells, to ensure that you are able to target and inject the heart. To practice with Evans blue dye, dilute it to 1 % in PBS, and inject an equivalent volume to the volume you plan to use for the stem cell injection. Perform the following protocol, substituting Evans blue dye for the stem cells, and euthanize experimental animals immediately after the procedure. Open the chest cavity and make an incision through the heart to determine whether the dye was injected into the cardiac muscle or into the chest cavity.

Alternatively, the optimal procedure for accurately injecting cells into the heart without opening the chest cavity would be to use ultrasound-guided injections. Specialized cardiac ultrasound equipment for use in rodents, such as the high-resolution VEVO2100 in vivo ultrasound imaging system from VisualSonics, is ideal for this method.

1. Anesthetize mice as in Section 3.7, step 1. Keep cells for injection in a sterile Eppendorf tube on ice until the injection.

2. Place the mouse on its back and secure the forepaws of the mouse to the surface of the table using laboratory tape.

3. Scrub down the chest area with Bactoshield CHG 4 % solution (STERIS, MO) in preparation for the injection. Draw up 5×10^6 cells in 50 µL saline into an insulin syringe (3/10 cc with a 29 G needle).

4. Use a forefinger to define the left ribcage. The needle will be inserted between the fourth and fifth ribs to inject cells into the left ventricle. This position also allows the needle of insulin syringe to penetrate the intercostal muscle without hitting sternum and ribs (Fig. 5).

5. Position the needle approximately 0.5 mm to the right side of sternum (Fig. 5). Insert the needle in the second intercostal space, approximately 3 mm to the left of the vertebral column until a resistance to the forward motion of the needle is felt. Insertion of the needle into the chamber of the heart will be verified by a decrease in resistance as well as the presence of a flashback of blood in the barrel of the syringe. Insertion of the needle into the wall of the heart will be verified by the absence of a flashback of blood, and movement of the heart. Inject cells slowly and retract the needle gently as soon as the cells have been delivered to avoid causing cardiac damage with the needle.

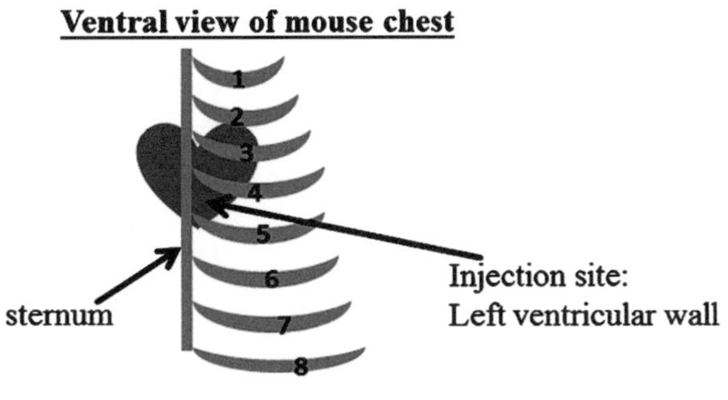

Fig. 5 Schematic of the mouse rib cage and site for intracardiac injection of stem cells. When targeting the left ventricle for injection of cells, insert the needle between the fourth and fifth ribs

6. After injection of cells, withdraw the nosecone and place the mouse on a cloth on top of a heating pad and monitor it until it is fully recovered.

7. For postprocedural pain management following intracardiac injection or systemic injection of cells, provide mice with 0.07 mg/kg of buprenorphine subcutaneously every 6–12 h as needed (increased activity, hunched posture, or a reluctance to move are indications that mice are in pain or distress and that buprenorphine is required to alleviate discomfort).

3.7 Detection of Stem Cells Using MRI

Acquire images of each mouse before and after stem cell injection to account for any variation among individual mice. Injection of unlabeled stem cells, or an equivalent volume of saline solution, is an appropriate control for specificity of detection of the SPIO-labeled cells. When injecting cells into the gastrocnemius, injection of saline solution only into the contralateral gastrocnemius serves as a helpful control for establishing changes in contrast specifically related to SPIO-labeled cells and distinguishing them from changes in contrast due to disturbances in the tissue resulting from the injection procedure (Fig. 4). Place a phantom in the holder next to the leg injected with stem cells to distinguish it from the leg injected with saline solution (Fig. 4i–l).

1. Anesthetize mice in the induction chamber of the anesthesia unit with isoflurane at approximately 5 %, waiting until the mouse is completely immobile.

2. Move the mouse into the holder (Fig. 1) (see Note 3), and switch the concentration of Isoflurane from 5 % in the chamber to 1.5–2.0 % at the holder. A proper oxygen flow rate is essential and was kept at 1.5 l/min. Ensure that warm air is passing through the holder during entire MRI session to keep the temperature at a physiologically comfortable range (37 ± 1 °C) (see Note 4).

3. Place the holder in the RF coil (Fig. 1) and then place the coil into the magnet, orienting the mouse in an upright position.

4. Start the MRI with RF coil tuning/matching by manual adjustment. Acquire an NMR signal (with no gradients) from the whole body and shim following Vnmrj (Varian/Agilent) software protocol. A conventional single pulse (SPULS) RF sequence can be employed. To avoid saturation a very small flip angle (5–10°) should be used. An offset radiofrequency should be found and the RF coil should be calibrated routinely. The same procedure should be repeated after any FOV changes.

5. Shimming. In contrast to high-resolution NMR spectroscopy, adjustment of the static magnetic field B_0 homogeneity, commonly known as *shimming* is not essential in MRI experiment because of strong inhomogeneous gradient pulses application.

However, shimming improves image quality. In our experiments the NMR signal for whole body of the mouse was shimmed manually with the best NMR linewidth around 200 Hz. For a particular area of interest (AOI), first- and second-order shims can be adjusted with FASTMAP automated shimming (13).

6. To ensure that the mouse is appropriately positioned, a series of quick low-resolution images should be obtained in the transverse, sagittal, and coronal planes. Quick images may be obtained using conventional SCOUT or Fast Spin-Echo Multi-Slice (FSEMS) pulse sequences. SCOUT is a variation of the gradient echo protocol that can provide quick images with good T_2^* contrast. FSEMS is well suited for obtaining good anatomical images in a short scan time. We frequently used the FSEMS protocol because it provided better image quality (Fig. 4).

7. After ensuring that the mouse is properly positioned, high-resolution T_2-weighted images may be obtained. A high image quality Spin-Echo-Multi-Slice (SEMS) protocol can be used with the following MRI acquisition parameters: slice thickness of 0.5 mm, repetition time (TR) of 2 s, and a matrix size of 256×256. The echo time (TE) may be varied between 9.5 and 50 ms to obtain an image with good T_2-weighted contrast of SPIO nanoparticles and high anatomical resolution. The FOV may be adjusted from 60 mm, to acquire an image of the entire mouse body, to 25 mm for acquisition of images focused on a particular area of the body. Receiver gain and radiofrequency transmit power should be optimized utilizing Vnmri software options. Using the tunable radiofrequency coil (5), the average scan time for high-resolution images of the whole body of mice was 18 min (4). Images may be processed with Matlab codes or VnmrJ software (Varian/Agilent). In T_2-weighted images, SPIO-labeled nanoparticles appear as hypointense regions (Fig. 4). See Note 11 regarding potential localization of stem cells to the lungs following systemic injection, and detection of cells in the lungs with MRI.

8. Analysis of postmortem tissue should also be performed to verify the presence of iron-labeled cells in tissue.

4 Notes

1. Isolation of ADM is detailed in Berry et al. (10). Protocols for differentiation of ADMs into myelinating glial cells in vitro are provided in Wang et al. (16).

2. The RF coil is one of the most important elements of any MRI/NMR system. Coil design and characteristics directly affect image quality. The RF coil used for our studies

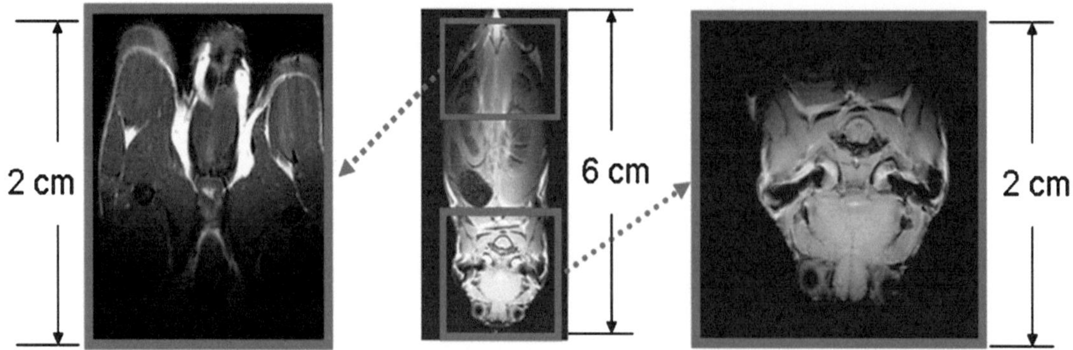

Fig. 6 Whole body imaging, as well as enhanced focusing on a particular area of interest (AOI) is useful for stem cell tracking in vivo. The field of view (FOV) adjustable radiofrequency (RF) coil can be used to localize stem cells following injection, and then adjusted to provide enhanced, high-resolution images of a particular region of tissue containing stem cells

(4, 5, 14, 15) overcomes limitations that occur in MRI/NMR measurements of comparatively large subjects with high dielectric (and other) losses (water, living tissues, etc.) as a result of the restricted tuning/matching range. The coil design expands the tuning range of resonance frequency from the 1–3 % typical for commercial coils up to 30 %, maintaining an appropriate impedance match. Coil electrical symmetry is preserved at the whole tuning range, resulting in high coil Q-factor and RF field homogeneity inside the coil. To improve image resolution in a particular AOI of a subject, an option of FOV adjustment was introduced into RF coil construction. The FOV-adjustable RF coil allows whole body scans after cell engraftment to detect injected cells in various organs, followed by high-resolution scans of the tissues where the cells were found without any visible "alias" artifacts and with high spatial resolution (Fig. 6). FOV adjustment also allows significantly decreased imaging acquisition time ($t \approx FOV^2$).

3. Limitations in space and unusual subject positioning at high field vertical bore magnet MRI system make holder design and construction particularly important. The holder should support an immobilized mouse in a vertical position during the MRI session inside a narrow (3.0 cm inner diameter) RF coil. The ultra-high magnetic field MRI holder design (Fig. 1) ensures safety and comfort for experimental animals. The flow of the Oxygen/Isoflurane mixture is directed to the nose of the mice to increase anesthesia efficiency (Fig. 1). The Oxygen/Isoflurane mixture is directed through a separate region of the holder to allow air flow to the mouse and subsequently out through an air filter that releases safe air into the room.

4. It is optimal to ensure that the warm air and Oxygen/Isoflurane flow pass through separate channels (Fig. 1) to prevent

interference with each other. The gas delivery schematic allows for controlled anesthesia for the duration of the MRI session.

5. Injection of excess DiI will label host tissue and may therefore result in false positive labeling. The repeated washes also serve to remove fetal bovine serum from the solution, which may trigger an immune reaction when injecting the cell solution into mice.

6. It is important to ensure a high rate of cell labeling (>95 %) with DiI prior to injection, to identify donor cells in host tissue for analysis of donor cell fate (Fig. 2), as well as determining donor cell localization and migration after injection. If the percentage of cells labeled with DiI prior to injection is low, it may be difficult to detect donor cells after injection. This will lead to uncertainty regarding cell survival after injection, and it may not be possible to correlate the donor cells with potential functional differences in the host after stem cell injection. In addition, DiI labeling will decrease with cell proliferation. Therefore limited proliferation of cells in host tissue after injection, particularly in cases where the donor cell population was poorly labeled prior to injection, may further deplete DiI, making detection of donor cells in host tissue difficult or impossible. It may be important to determine whether DiI-labeled donor cells proliferate after injection, particularly if it is difficult to detect DiI-labeled stem cells several days following injection. This can be done using an antibody to Ki67, which is present in all phases of the cell cycle of actively dividing cells, to determine whether DiI-labeled cells express Ki67 at various times following injection. Alternatively, donor cells can be labeled with BrdU prior to injection, and an antibody to BrdU can be used on host tissue after injection to detect donor cells. As a cautionary note, if using BrdU to label stem cells that are also colabeled with DiI, be aware that a low level of HCl is necessary to partially denature DNA in postmortem tissue to allow binding of BrdU antibody to BrdU residues in the donor cell DNA. This incubation with HCl will cause loss of DiI from stem cell membranes. For this reason, we prefer to fix postmortem tissue with 3.7 % formaldehyde and use an antibody to Ki67 to detect division of DiI-labeled cells.

7. We have used FeREX to label ADMs at concentrations identical to Feridex, and have found that the ADM readily incorporate it and will exhibit morphology changes consistent with differentiation into myelinating glial cells, as they also did following labeling with Feridex (4) (Fig. 3b), indicating that FeREX is a good substitute for Feridex.

8. Many studies have been conducted to optimize the use of transfection agents for enhancing labeling of some cell

populations with SPIO. Frank et al. have explored the use of commercially available transfection reagents for this purpose (17, 18), testing a variety of agents on different cell types including stem cell and precursor cell populations such as mesenchymal stem cells, oligodendrocyte progenitor cells.

9. When anesthetizing mice that are older, or in poor health, such as the mdx/utrn−/− mouse model for Duchenne muscular dystrophy (4, 10–12), inhalation anesthesia is a more gentle method. For example, the mdx/utrn−/− mice do not awaken from anesthesia induced with xylazine/ketamine mixtures, of which we have tried different ratios and concentrations.

10. Pass cells through the needle to be used for injection, and then culture the cells after passage through the needle. This is particularly important for small gauge needles such as insulin syringes with 28 G needles to ensure that the cells remain viable afterwards. Large cells may not survive passage through small needles.

11. Systemically injected ADM localize to the lungs (4), as do many other stem cells when delivered systemically. It is difficult to use T2-weighted imaging to visualize iron oxide-labeled stem cells in the lungs because of the low water content and constant motion. However, the use of respiratory and cardiac gating, together with specialized gases, may be used to overcome this difficulty (19–21). Given the difficulty in detection and requirement for specialized equipment to detect stem cells labeled with SPIO nanoparticles in the lungs, performing addition injections of stem cells labeled with fluorescent dyes such as DiI are helpful to visualize the cells in postmortem tissues to verify whether donor cells migrate to, or become trapped in, the lungs.

References

1. Rogers W, Meyer C, Kramer C (2006) Technology insight: in vivo cell tracking by use of MRI. Nat Clin Pract Cardiovasc Med 3:554–562

2. Waters E, Wickline S (2008) Contrast agents for MRI. Basic Res Cardiol 103:114–121

3. Bowen C, Zhang X, Saab G et al (2002) Application of the static dephasing regime theory to uperparamagnetic iron-oxide loaded cells. Magn Reson Med 48:52–61

4. Odintsov B, Chun J, Mulligan J et al (2011) 14.1 T whole body MRI for detection of mesoangioblast stem cells in a murine model of Duchenne muscular dystrophy. Magn Reson Med 66:1704–1714

5. Odintsov B (2011) Tunable radiofrequency coil. US Patent US 2010/10013483 A1, 1 Nov 2011

6. Doty F, Entzminger G, Kulkarni J et al (2007) Radio frequency coil technology for small-animal MRI. NMR Biomed 20:304–345

7. Jing X, Yang L, Duan X et al (2008) In vivo MR imaging tracking of magnetic iron oxide nanoparticle labeled, engineered, autologous bone marrow mesenchymal stem cells following intra-articular injection. Joint Bone Spine 75:432–438

8. Hoehn M, Kustermann E, Blunk J et al (2002) Monitoring of implanted stem cell migration in vivo: a highly resolved in vivo magnetic resonance imaging investigation of experimental

stroke in rat. Proc Natl Acad Sci U S A 99:16267–16272

9. Syková E, Jendelová P, Herynek V (2009) MR tracking of stem cells in living recipients. Methods Mol Biol 549:197–215

10. Berry S, Liu J, Chaney E et al (2007) Multipotential ADM stem cell therapy in the mdx/utrn$^{-/-}$ mouse model for Duchenne muscular dystrophy. Regen Med 2:275–288

11. Grady R, Teng H, Nichol M et al (1997) Skeletal and cardiac myopathies in mice lacking utrophin and dystrophin: a model for Duchenne muscular dystrophy. Cell 90:729–738

12. Deconinck A, Rafael J, Skinner J (1997) Utrophin-dystrophin-deficient mice as a model for Duchenne muscular dystrophy. Cell 90:717–727

13. Gruetter R, Tkáč I (2000) Field mapping without reference scan using asymmetric echo-planar techniques. Magn Reson Med 43:319–323

14. Brozoski T, Odintsov B, Bauer C (2012) Gamma-aminobutyric acid and glutamic acid levels in the auditory pathway of rats with chronic tinnitus: a direct determination using high resolution point-resolved proton magnetic resonance spectroscopy (1H-MRS). Front Syst Neurosci 6:9

15. Byung H, Cho S, Park J et al (2010) Synthetic dimyristoylphosphatidylcholine liposomes assimilating into high-density lipoprotein promote regression of atherosclerotic lesions in cholesterol-fed rabbits. Exp Biol Med 235:1194–1203

16. Wang L, Kamath A, Frye J et al (2012) Aorta-derived ADMs differentiate into oligodendrocytes by inhibition of the Rho kinase signaling pathway. Stem Cells Dev 21 (7):1069–1089

17. Frank J, Miller B, Arbab A et al (2003) Clinically applicable labeling of mammalian and stem cells by combining superparamagnetic iron oxides and transfection agents. Radiology 228:480–487

18. Frank J, Zywicke H, Jordan E et al (2002) Magnetic intracellular labeling of mammalian cells by combining (FDA-approved) superparamagnetic iron oxide MR contrast agents and commonly used transfection agents. Acad Radiol Suppl 2:S484–S487

19. Driehuys B, Hedlund L (2007) Imaging techniques for small animal models of pulmonary disease: MR microscopy. Toxicol Pathol 35:49–58

20. Driehuys B, Nouls J, Badea A et al (2008) Small animal imaging with magnetic resonance microscopy. ILAR J 49:35–53

21. Loebinger M, Kyrtatos P, Turmaine M et al (2009) Magnetic resonance imaging of mesenchymal stem cells homing to pulmonary metastases using biocompatible magnetic nanoparticles. Cancer Res 69:8862–8867

Methods Molecular Biology (2013) 1052: 195–201
DOI 10.1007/7651_2013_27
© Springer Science+Business Media New York 2013
Published online: 3 May 2013

Molecular Imaging and Tracking Stem Cells in Neurosciences

Toma Spiriev, Nora Sandu, and Bernhard Schaller

Abstract

Stem cell transplantation is a promising new therapeutic option in different neurological diseases. However, it is not yet possible to translate its potential from animal models to clinical application. One of the main problems of applying stem cell transplantation in clinical medium is the difficulty of detection, localization, and examination of the stem cells in vivo at both cellular and molecular levels. State-of-the-art molecular imaging techniques provide new and better means for noninvasive, repeated, and quantitative tracking of stem cell implant or transplant. From initial deposition to the survival, migration, and differentiation of the transplant/implanted stem cells, current molecular imaging methods allow monitoring of the infused cells in the same live recipient over time. The present review briefly summarizes and compares these molecular imaging methods for cell labeling and imaging in animal models as well as in clinical application and sheds light on consecutive new therapeutic options if appropriate.

Keywords: Molecular imaging, Stem cells, Neuroscience, Animal model

1 Introduction

Stem cell transplantation is a promising new therapeutic option in different neurological diseases (1–6). Such cell therapies target towards CNS injuries where neurogenesis is thought to play a special key role in treatment of stroke, and developmental or neurodegenerative disorders, or even after brain surgery (2, 3, 7–11). Therefore, different cell types of stem cells are under evaluation for their therapeutic efficacy in cell-based therapies of neurological disorders (1–3, 7, 12–15). Under these circumstances, it is critically important to track the location of directly transplanted or infused cells that can serve as gene carrier/delivery vehicles for the treatment of disease processes and be able to noninvasively monitor the temporal and especially spatial homing of these cells to target tissues (9, 11).

Over the last few years, our understanding of the behavior of stem cells is mostly gained from in vitro studies. However, it is not yet entirely possible to translate its potential from animal models to clinical application. For most of the laboratory investigations to truly address the clinic, either large animal models or patients should be used in studies. In many pathophysiological states, large animal models have been shown to be similar to human in respect to disease progression (16). However, before these models are routinely applied clinically, there are a number of questions, such as dose, timing of delivery, homing, etc., that need to be answered.

One of the main problems of applying stem cell transplantation in clinical medium is the difficulty of detection, localization, and examination of the stem cells in vivo at both cellular and molecular levels. In addition, after cells are transplanted into the living subject, it becomes critical to understand the exact biology of transplanted cells and the interaction with the microenvironment (14, 17). These issues include finding appropriate probes that elicit minimal or no immunogenic response, enhancing transfection stability, and minimizing the potential interference of the stem cell function and differentiation from vector transfection or transduction (10, 11, 18–20).

2   Molecular Imaging Modalities Using Magnetic Resonance Imaging

Although magnetic resonance imaging (MRI) has been extensively used to track cells in vivo (7, 8, 21), positron emission tomography (PET) seems to be more promising for long-term in vivo monitoring of transplanted cells by MI (7–9, 14, 18, 20, 21). This superiority is due to the higher specificity of PET and its ligands as well as its ability to detect reporter genes (9) (Table 1).

MRI technique only reflects biological activity of the turnover indirectly by detecting the changes of the blood–brain barrier. While visualization by MRI has a high spatial resolution, cells

Table 1

Overview about molecular imaging systems in animals (adapted from (11, 21))

Technique	Resolution	Time	Imaging agents	Target
MR	10–100 mm	Minutes–hours	Gadolinium, dysprosium iron oxide particles	A, P, M
PET	1–2 mm	Minutes	18F, 11C, 15O	P, M
BLI	Several mm	Minutes	Luciferins	M

MR magnetic resonance, *PET* positron emission tomography, *BLI* bioluminescence imaging, *A* anatomical, *P* physiological, *M* molecular

need to be loaded with contrast agents, like superparamagnetic iron oxides, which might display some toxicity towards the implanted cells and surrounding CNS tissue (7, 8, 22). Another disadvantage of these contrast agents is their leakage out of necrotic cells and their consequent uptake by endogenous cells. This might result in the false identification of recipient-site cells as implant cells and cause unreliable survival and localization outcomes.

Therefore, tagging the stem cells with different ligands by PET can turn these cells into interesting probes for different MI modalities (22). The advantages of such methods are the high detection sensitivity of PET imaging techniques and possible immediate translation to clinical practice, as most of the used radiotracers are already in clinical use. During the last few years, such translation from bench to bedside was forced.

3 Molecular Imaging Modalities Using Positron Emission Tomography

The ideal MI modality is one that has excellent spatial resolution and cell detection sensitivity, can guide the delivery of cells, and can serially follow stem cells and their fate.

Currently, no such imaging modality exists. As already mentioned above, every MI modality should be chosen depending on the question that is being asked.

The central challenge for MI is to develop specific reporter probes that come in many shapes and sizes, but their principal components are a targeting molecule and a specific ligand. In vivo, the probe should be targeting and visualizing the biologic process of interest. Many types of MI probes have been developed for different MI techniques (22–24). Popular target choices are receptors, enzymes, and cytokines, which are often expressed at higher level or differently in certain CNS pathologies such as leucine-rich repeat kinase 2 (LRRK2) and alpha-synuclein (SNCA) in Parkinson disease (6, 15) or apolipoprotein E in Alzheimer disease (24, 25). Natural ligands, such as receptor agonists, can be used, or specific ligands, such as antibodies or peptides, can be developed (20). However, instead of visualizing reporter molecules, it is also possible to label intact cells, allowing cell tracking and gaining knowledge on cell behavior, such as stem cell survival following transplantation.

Rueger et al. (18) used PET and the radiotracer 3′-deoxy-3′-[(18)F] fluoro-L-thymidine ([(18)F]FLT) that enables MI and measuring of proliferation to noninvasively detect endogenous neural stem cells in the normal and diseased adult rat brain in vivo. They could visualize neural stem cell niches in the living rat brain, identified as increased [(18)F] FLT binding in the subventricular zone and the hippocampus (18). Furthermore, they could quantify neural stem cell mobilization caused by pharmacological stimulation or by focal cerebral ischemia (18). Such monitoring by MI is also of great clinical importance and should therefore cover two aspects: visualization and quantization of cell migration, as well as functional status. Moreover, although it cannot generally provide functional activation data that would also afford the investigation of downstream effects of regional activity (i.e., functional circuitry assessment), PET is currently the only reliable option to provide specific and quantitative data for the presence of receptors or molecules in a particular region. It is therefore currently the MI modality of choice to assess the presence of dopaminergic grafts in vivo (26).

Another important aspect of MI is that one could monitor gene expression, track cells in normal and abnormal development, map dynamic protein interactions, and check cell transplantation therapy. Taking this first step further, one could follow the effects of gene therapy, in which stem cells are genetically modified to produce a therapeutic effect.

Reporter gene imaging is a technique in which gene products (i.e., reporter proteins) are imaged in vivo (27). Essentially, a reporter gene is transcribed to mRNA, which in turn is translated into a reporter protein (which are far more abundant in the cell than DNA or RNA). A good reporter protein must be easy to assay and must not normally be expressed in the cells of interest or, when encoding for endogenous proteins, must be expressed at much higher levels than normal. Thus reporter gene imaging currently represents a powerful approach to study the physiology and biology of transplanted cells in vivo. Evidently, cells should remain viable and functional after the labeling procedure. Cellular contrast agents should ideally remain within the desired cell type and not dilute with cell division, to enable reliable longitudinal studies.

The gene of interest is unlikely to encode by MI to a visible protein, though the protein of interest may interact with exogenous reporter molecules. Often, the gene of interest is teamed up with a reporter gene (27–29). These genes can be engineered so that they are both driven by the same promoter. On activation of this promoter (which can be conditional or tissue specific), the expression of both genes is simultaneously enhanced; MI, the reporter protein, thus "reports" on the expression of the gene of interest. For the application for stem cell monitoring, the reporter gene is incorporated into the cell before cell transplantation into the living subject. If the stem cells are viable (e.g., after transplantation), the reporter gene will be expressed and the protein (e.g., enzyme, cell surface, receptor) will be encoded. Such reporter gene MI is currently in a preclinical stage, but its potential seems to be enormous (20).

4 Molecular Imaging Modalities Using Bioluminescence Imaging

Bioluminescence imaging (BLI) has more widely used applications in molecular imaging of stem cells, either alone or in combination with other state-of-the-art imaging methods, because of its noninvasive molecular and cellular level detection ability, high sensitivity, and low cost in comparison with other imaging technologies (30, 31). However, BLI cannot present the accurate location and intensity of the inner bioluminescent sources. Bioluminescent tomography (BLT), however, shows its advantages in determining the bioluminescent source distribution inside a small animal or phantom (32, 33). By utilizing CT or even MRI information acquired by an X-ray detector, the three-dimensional location can be reconstructed using some BLT reconstruction methods such as the adaptive finite element method and Bayesian method. These tomography imaging methods can be used in detection of the target tumor cells and assist in the diagnosis and evaluation of treatment efficacy more accurately. Furthermore, the clinical application of BLI needs to develop some novel probes which can be used in humans.

5 Some Problems Associated with Positron Emission Tomography Molecular Imaging

However, several problems must be solved before reporter gene imaging can fully and safely be applied clinically. These issues include finding appropriate probes that elicit minimal or no immunogenic response, enhancing transfection stability, and minimizing the potential interference of the stem cell function and differentiation from vector transfection or transduction (11, 20, 28).

On one side, PET has significant flexibility for the production of specific probes for the detection of different processes in the living subject (almost any compound can be labeled with a radionuclide), which is a significant advantage as it allows the researcher to first identify the molecule that needs to be studied, and then design a specific probe that will target that molecule (8–10, 13, 18, 24). In relation with stem cell transplantation in neuroscience, PET is currently the method of choice, alone or in combination with MRI. However, the production of PET probes is complex, and needs advanced chemistry and very tight quality control. In addition, depending on the half-life of the radioisotope used, it requires an on-site (or at least nearby) cyclotron that limits this strategy to medium to large research centers. From the imaging standpoint, all electron–positron annihilations (whether from ^{18}F, ^{15}O, or ^{11}C) result in the production of photons of 511 KeV, and as such we cannot detect differences in registered signals (10, 34–36).

References

1. Arvidsson A, Collin T, Kirik D et al (2002) Neuronal replacement from endogenous precursors in the adult brain after stroke. Nat Med 8:963–970

2. Bang OY, Lee JS, Lee PH, Lee G (2005) Autologous mesenchymal stem celltransplantation in stroke patients. Ann Neurol 57:874–882

3. Karussis D, Karageorgiou C, Vaknin-Dembinsky A et al (2010) Safety and immunological effects of mesenchymal stem cell transplantation in patients with multiple sclerosis and amyotrophic lateral sclerosis. Arch Neurol 67:1187–1194

4. Lichtenwalnder RJ, Parent JM (2006) Adult neurogenesis and ischemic forebrain. J Cereb Blood Flow Metab 26:1–20

5. Martino G, Franklin RJM, Van Evercooren AB, Kerr DA (2010) Stem cell transplantation in multiple sclerosis: current status and future prospects. Nat Rev Neurol 6:247–255

6. Schaller B, Cornelius JF, Sandu N (2003) Molecular medicine successes in neurosciences. Mol Med 14:361–364

7. Arbab AS, Frank JA (2008) Cellular MRI and its role in stem cell therapy. Regen Med 3:199–215

8. Modo M, Hoehn M, Bulte J (2005) Cellular MR imaging. Mol Imaging 4:1–21

9. Schaller B (2004) Usefulness of positron emission tomography in diagnosis and treatment follow-up of brain tumors. Neurobiol Dis 15:437–448

10. Schaller BJ, Modo M, Buchfelder M (2007) Molecular imaging of brain tumors: a bridge between clinical and molecular medicine? Mol Imaging Biol 9:60–71

11. Sandu N, Momen-Heravi F, Sadr-Eshkevari P, Schaller B (2012) Molecular imaging for stem cell transplantation in neuroregenerative medicine. Neurodegener Dis 9:60–67

12. Bentz K, Molcanyi M, Hess S et al (2006) Neural differentiation of embryonic stem cells is induced by signalling from non-neural niche cells. Cell Physiol Biochem 18:275–286

13. Környei Z, Szlávik V, Szabó B, Gócza E, Czirók A, Madarász E (2005) Humoral and contact interactions in astroglia/stem cell co-cultures in the course of glia-induced neurogenesis. Glia 49:430–444

14. Nakano K, Migita M, Mochizuki H, Shimada T (2001) Differentiation of transplanted bone marrow cells in the adult mouse brain. Transplantation 71:1735–1740

15. Yu D, Silva GA (2008) Stem cell sources and therapeutic approaches for central nervous system and neural retinal disorders. Neurosurg Focus 24:E10

16. Bloor CM, White FC, Roth DM (1992) The pig as a model of myocardial ischemia and gradual coronary artery occlusion. In: Swindle M, Moody MM, Phillips DC, Ames LD (eds) Swine as models in biomedical research. Iowa State University Press, Iowa, pp 163–175

17. Hess DC, Abe T, Hill WD et al (2004) Hematopoietic origin of microglial and perivascular cells in brain. Exp Neurol 186:134–144

18. Rueger MA, Backes H, Walberer M et al (2010) Noninvasive imaging of endogenous neural stem cell mobilization in vivo using positron emission tomography. J Neurosci 30:6454–6460

19. Matusik E, Wajgt A, Janowska J et al (2009) Cell adhesion molecular markers in ischaemic stroke patients: correlation with clinical outcome and comparison with primary autoimmune disease. Arch Med Sci 5:182–189

20. Sandu N, Schaller B (2010) Stem Cell transplantation in brain tumors: a new field for molecular imaging. Mol Med 16:33–37

21. Rudin M, Weissleder R (2003) Molecular imaging in drug discovery and development. Nat Rev Drug Discov 2:123–131

22. Solanki A, Kim JD, Lee KB (2008) Nanotechnology for regenerative medicine: nanomaterials for stem cell imaging. Nanomedicine (Lond) 3:567–578

23. Schaller BJ, Buchfelder M (2006) Neuroprotection in primary brain tumors: sense or nonsense? Expert Rev Neurother 6:723–730

24. Schaller B (2008) State-of-the-art-imaging-methods to investigate the neurovascular mechanism in the origin of Alzheimer's disease. Differential diagnostic evaluations to other types of dementia. Neuropsychiatr Dis Treat 4:585–612

25. Yamagata K, Urakami K, Ikeda K et al (2001) High expression of apolipoprotein EmRNA in the brains with sporadic Alzheimer's disease. Dement Geriatr Cogn Disord 12:57–62

26. Yeh E, Gustafson K, Boulianne GL (1995) Green fluorescent protein as a vital marker and reporter of gene expression in Drosophila. Proc Natl Acad Sci U S A 92:7036–7040

27. Schaller BJ, Cornelius JF, Sandu N, Buchfelder M (2008) Molecular imaging of brain tumors: personal experience and review of the literature. Curr Mol Med 8:711–712

28. Joyce N, Annett G, Wirthlin L et al (2010) Mesenchymal stem cells for the treatment of neurodegenerative disease. Regen Med 5:933–946

29. Koehne G, Doubrovin M, Doubrovina E et al (2003) Serial in vivo imaging of targeted migration of human HSV-TK-transduced antigen-specific lymphocytes. Nat Biotechnol 21:405–413

30. Gould SJ, Subramani S (1988) Firefly luciferase as a tool in molecular and cell biology. Anal Biochem 175:5–13

31. Keenan TM, Nelson AD, Grinager JR, Thelen JC, Svendsen CN (2010) Real time imaging of human progenitor neurogenesis. PLoS One 5: e13187

32. Massoud TF, Singh A, Gambhir SS (2008) Noninvasive molecular neuroimaging using reporter genes: part I, principles revisited. AJNR Am J Neuroradiol 29:229–234

33. Tokuda T, Qureshi MM, Ardah MT et al (2010) Detection of elevated levels of {alpha}-synuclein oligomers in CSF from patients with Parkinson disease. Neurology 75:1766–1770

34. Gaura V, Bachoud-Levi A-C, Ribeiro M-J et al (2004) Striatal neural grafting improves cortical metabolism in Huntington's disease patients. Brain 127:65–72

35. Lee BD, Shin JH, VanKampen J et al (2010) Inhibitors of leucine-rich repeat kinase-2 protect against models of Parkinson's disease. Nat Med 16:998–1000

36. Ray P, Tsien R, Gambhir SS (2007) Construction and validation of improved triple fusion reporter gene vectors for molecular imaging of living subjects. Cancer Res 67:3085–3093

Methods Molecular Biology (2013) 1052: 203–215
DOI 10.1007/7651_2013_15
© Springer Science+Business Media New York 2013
Published online: 4 June 2013

Bioluminescence Imaging of Human Embryonic Stem Cell-Derived Endothelial Cells for Treatment of Myocardial Infarction

Weijun Su, Liang Leng, Zhongchao Han, Zuoxiang He, and Zongjin Li

Abstract

Myocardial infarction is a leading cause of mortality and morbidity worldwide, and current treatments fail to address the underlying scarring and cell loss, which is a major cause of heart failure after infarction. The novel strategy, therapeutic angiogenesis and/or vasculogenesis with endothelial progenitor cells transplantation holds great promise to increase blood flow in ischemic areas, thus rebuild the injured heart and reverse the heart failure. Given the potential of self-renewal and differentiation into virtually all cell types, human embryonic stem cells (hESCs) may provide an alternate source of therapeutic cells by allowing the derivation of large numbers of endothelial cells for therapeutic angiogenesis and/or vasculogenesis of ischemic heart diseases. Moreover, to fully understand the fate of implanted hESCs or hESC derivatives, investigators need to monitor the motility of cells in living animals over time. In this chapter, we describe the application of bioluminescence reporter gene imaging to track the transplanted hESC-derived endothelial cells for treatment of myocardial infarction. The technology of inducing endothelial cells from hESCs will also be discussed.

Keywords: Human embryonic stem cells, Endothelial cells, Myocardial infarction, Bioluminescence imaging, Firefly luciferase, Green fluorescent protein

1 Introduction

Due to obstruction of coronary vessels, myocardial infarction causes the injury and subsequent necrosis of myocardium, and therapeutic angiogenesis/vasculogenesis holds promise for the cure of those diseases (1). Endothelial progenitor cells transplantation intends to foster the formation of arterial collaterals and to promote the regeneration of damaged tissues, which offers a promising strategy to increase the blood flow in patients with ischemic heart disease (2, 3). However, the source of adult endothelial progenitor cells is limited, which highlights the need for a consistent and renewable source of endothelial cells for clinical applications. With their capacity for unlimited self-renewal and pluripotency, human embryonic stem cells (hESCs) may provide an alternate source of endothelial cells for novel transplantation therapies of ischemic diseases by supporting angiogenesis and vasculogenesis (4–6).

To thoroughly elucidate the therapeutic potential of hESC derivatives, it is significant for investigators to monitor their biological behavior after transplantation in vivo. Comparing with traditional technologies, which need to sacrifice the animals at the endpoint of experiments, bioluminescence imaging (BLI) helps investigators achieve the noninvasive and longitudinal visualization of spatio-temporal kinetics of transplanted cells. In our laboratory, we use a lentiviral vector carrying ubiquitin promoter which drives double fusion (DF) construct containing enhanced green fluorescent protein (GFP) and Firefly luciferase (Fluc) reporter gene (7, 8). Fluc synthesized by target cells can interact with exogenously administered D-luciferin, emitting visible light which can be detected by a cool-charged couple device (CCD) camera, thus achieve localization of the transplanted cells (Fig. 1). In addition, GFP expression can be used to select a nearly homogeneous population of cells with fluorescence activated cell sorting (FACS) and

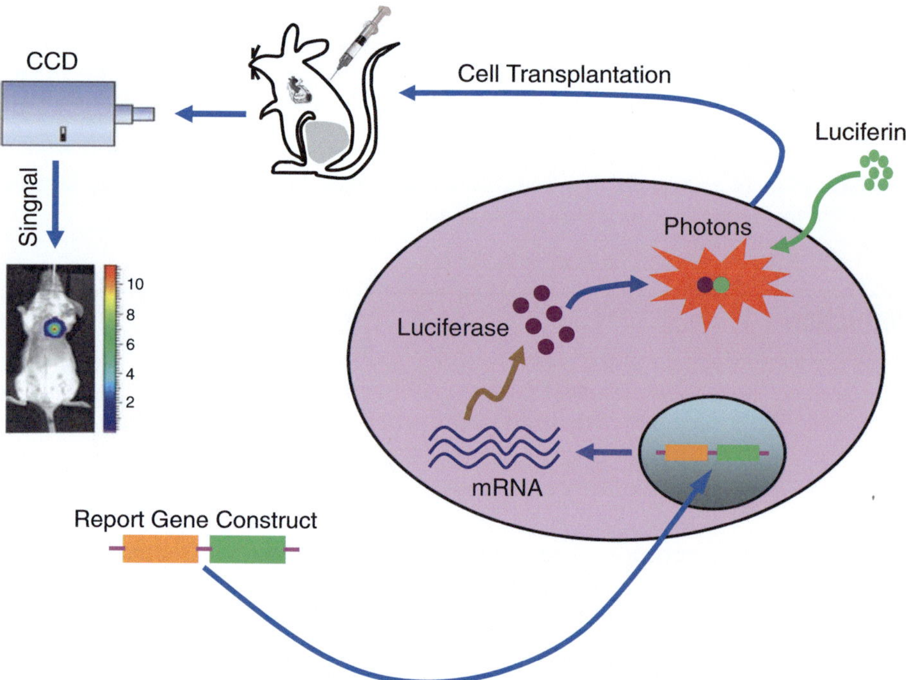

Fig. 1 Conceptual basis of tracking the survival of transplanted stem cells in living animals by bioluminescence imaging (BLI). hESC was transfected with report gene, can produce luciferase. In the presence of luciferin, the cell can emit photons and can be detected by high-sensitive CCD camera. After hESC-ECs transplantation for ischemia heart model, cell surviving can be tracked by BLI. Reproduced with permission from ref. (1)

for evaluation of the graft site of transplanted cells at the endpoint by immunohistology. In this chapter, we label hESCs with DF reporter gene and monitor the hESC-derived endothelial cells (hESC-ECs) expressing reporter gene after transplanted intramyocardially in myocardial infarction experimental models by BLI.

2 Materials

2.1 Maintenance of Undifferentiated hESCs

1. BD Matrigel™ hESC-qualified Matrix (BD Biosciences, San Jose, CA, USA)
2. Knockout™ DMEM (Invitrogen, Carlsbad, CA, USA)
3. mTeSR1 medium (Stem Cell Technologies, Vancouver, BC, Canada)
4. Dulbecco's Phosphate-Buffered Saline (DPBS) (Invitrogen, Carlsbad, CA, USA)
5. Dispase (Invitrogen, Carlsbad, CA, USA)
6. Tissue culture dish (BD Falcon™, San Jose, CA, USA)
7. Corning® cell lifter (USA)
8. 100 % Ethanol
9. Microscope

2.2 Lentiviral Transduction of hESCs with Reporter Gene

1. Dulbecco's Modified Eagle Medium (DMEM), high glucose (Invitrogen, Carlsbad, CA, USA)
2. Fetal Bovine Serum (Invitrogen, Carlsbad, CA, USA)
3. Nonessential Amino Acids (NEAA) (Invitrogen, Carlsbad, CA, USA)
4. L-glutamine (Invitrogen, Carlsbad, CA, USA)
5. Sodium Pyruvate (Invitrogen, Carlsbad, CA, USA)
6. Opti-MEM medium (Invitrogen, Carlsbad, CA, USA)
7. Lipofectamine-2000 (Invitrogen, Carlsbad, CA, USA)
8. Polybrene (Millipore, Billerica, MA, USA)

2.3 Endothelial Differentiation of hESCs

1. Isove's modified Dulbecco's medium (IMDM) (Invitrogen, Carlsbad, CA, USA)
2. Knockout™ Serum Replacement (KSR) (Invitrogen, Carlsbad, CA, USA)
3. BIT (Invitrogen, Carlsbad, CA, USA)
4. Monothioglycerol (Sigma-Aldrich, USA, USA)
5. Penicillin–streptomycin (Invitrogen, Carlsbad, CA, USA)
6. bFGF (R&D Systems, Minneapolis, MN, USA)
7. VEGF (R&D Systems, Minneapolis, MN, USA)

8. Ultralow attachment dish (Corning Incorporated, Corning, NY, USA)

9. EBM-2 basal medium (Lonza, Switzerland)

10. EGM-2 Bullet Kit (Lonza, Switzerland)

11. Liberase Blendzyme IV (Roche, CA, USA)

12. 40-mm cell strainer (BD Biosciences, San Jose, CA, USA)

13. PE Mouse anti-human CD31 (BD Biosciences, San Jose, CA, USA)

14. Fibronectin (Invitrogen, Carlsbad, CA, USA)

15. 0.25 % trypsin–EDTA (Invitrogen, Carlsbad, CA, USA)

2.4 Myocardial Infarction Model and Cell Transplantation

1. Isoflurane (Forane, Shanghai, China)

2. Small animal respirator (Harvard Apparatus, Holliston, MA, USA)

3. Polypropylene suture (Ethicon, Auneau, France)

4. Animal hair clipper (Harvard Apparatus, size 40 clipper blade, Holliston, MA, USA)

5. Small scissors, large scissors, and forceps (FST, Foster City, CA, USA)

6. Hamilton syringe with a 30 G beveled needle (Hamilton Co, Bonaduz, GR, Switzerland)

2.5 BLI of Transplanted hESC-ECs In Vivo

1. Xenogen IVIS machine with portable anesthesia system (Caliper, Hopkinton, MA, USA)

2. D-luciferin Potassium Salt (Caliper, Hopkinton, MA, USA)

3. 25 g 1/2″ needle + 1 ml (cc) syringe (BD Biosciences, San Jose, CA, USA)

3 Methods

3.1 Maintenance of Undifferentiated hESCs

3.1.1 Coating Tissue Culture Dishes with Matrigel hESC-Qualified Matrix

1. Take out an aliquot of frozen BD Matrigel™ hESC-qualified Matrix from −20 °C refrigerator. Put it on ice and wait until liquid.

2. Dilute the thawed BD Matrigel™ Matrix 1:20 in cold DMEM, and mix it thoroughly.

3. Add the BD Matrigel™ Matrix solution to culture dishes and shake the dishes slightly to spread the Matrix solution around. For 100 mm dishes, 5 ml solution should be used.

4. Leave the coated dishes at 4 °C overnight.

5. Remove the BD Matrigel™ Matrix solution from coated dishes before use.

3.1.2 Thawing and Seeding of hESCs on Coated Plates in mTeSR1 Medium

1. Remove one hESCs vial from the liquid nitrogen storage tank using metal forceps. Immerse the vial in a 37 °C water bath without submerging the cap. Swirl the vial gently.

2. Take out the cryovial from water bath when only an ice crystal remains.

3. Immerse the vial into a 100 % ethanol bath to sterilize the outside of the tube. Briefly air-dry the vial in culture hood.

4. Transfer the content into a 15 ml conical tube with a 1 ml glass pipette and add 4 ml mTeSR1 medium into the tube, gently move the tube back and forth to mix the hESCs. Centrifuge the cells at 200 × g for 5 min.

5. Suck out the supernatant with pipette. Add 12 ml fresh mTeSR1 medium into the tube, and resuspend the cell pellet gently.

6. Transfer the cell suspension to BD Matrigel™ hESC-qualified Matrix-coated 10 mm dishes and shake the dishes side-to-side to spread the cells evenly.

7. Place the dish into incubator, and culture the cells at 37 °C, with 5 % CO_2. Check the undifferentiated state and change medium every day.

3.1.3 Passaging of hESCs

1. Remove the culture medium from the dish and rinse the cells with sterile DPBS.

2. Add 1 mg/ml dispase to the cells. For 100 mm dishes, 6 ml dispase per dish is recommended. Incubate for 5–7 min at 37 °C until the edge of colonies begins to detach from the dish (Note 1).

3. Aspirate the dispase from the dish and wash the cells twice with DPBS very gently.

4. Add 5 ml pre-warmed mTeSR1 medium into the dish and scrape the cells with a cell lifter. Collect the cell suspension into a 15 ml conical tube, and centrifuge it at 200 × g for 5 min.

5. Aspirate the supernatant and resuspend the cell pellet with fresh mTeSR1 medium.

6. Transfer the cell suspension into pre-coated 10 mm dishes at split ratio 1:3.

3.2 Lentiviral Transduction of hESCs with Reporter Gene

For delivery of the reporter gene, our group has employed a double fusion (DF) construct containing enhanced GFP and Fluc as reported (4, 9) (Fig. 2).

3.2.1 Production of Self-Inactivating Lentivirus

1. Thaw a vial of 293 T cells and culture the cells with DMEM medium, supplemented with 10 % FBS, 0.1 mM nonessential amino acids, 2 mM L-glutamine and 1 mM sodium pyruvate (Note 2).

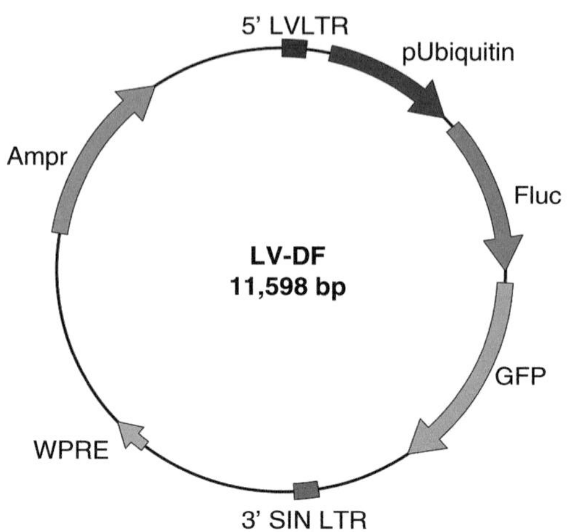

Fig. 2 The DF construct contains enhanced green fluorescent protein (GFP) and firefly luciferase (Fluc) reporter genes linked by a 5-amino acid linker (GSHGD)

2. Keep 293 T cells growing exponentially. On the day prior to transfection, seed 8×10^6 cells per 10 mm dish.

3. About 24 h after seeding, the cells reach nearly 90 % confluence. Before adding transfection reagents, aspirate medium and add 5 ml fresh culture medium to each dish.

4. Mix 15 μg DF plasmid (containing the bi-fusion construct), 10 μg HIV-1 packaging vector (pCMVDR8.2) and 5 μg vesicular stomatitis virus G glycoprotein-pseudotyped envelop vector (pMD.G) in 500 μl Opti-MEM medium. Add 30 μl lipofectamine-2000 to 500 μl Opti-MEM medium into another tube. The dosage of each packaging plasmid and lipofectamine-2000 should be optimized according to the lentiviral packaging system (Note 3).

5. Combine the plasmids and lipofectamine-2000 in 5 min. Wait for another 20 min and add the mixture to each dish.

6. 15–16 h after transfection, change the medium with 10 ml fresh complete DMEM medium (Note 4).

7. 24 h later, harvest the lentivirus-containing supernatant.

8. Centrifuge the supernatant at low speed ($200 \times g$ for 5 min), and purify it by passing through a 0.45 μm filter.

9. Concentrate the lentivirus by sediment centrifugation of the medium with an SW29 rotor at $50,000 \times g$ for 2 h.

10. After centrifugation, dissolve the viral pellets in 100 μl serum-free medium and store at $-80\ °C$ (Note 5).

3.2.2 Transduction of hESCs with Lentiviral Vector

1. hESCs cultured in Matrigel matrix pre-coated 6-well plates can be transduced 3–5 days after passage (Note 6).

2. On the day of infection, take out an aliquot of frozen virus from −80 °C refrigerator. Put it at room temperature and wait until liquid.

3. Add in total volume of 1,000 μl Opti-MEM (6-well plate) and 20 μl virus with polybrene (8 μg/ml) to cells and incubate for 2–4 h (Note 7).

4. Wash off the virus with PBS and then add 3 ml fresh mTeSR1 medium.

5. Incubate for 48–72 h then assay for gene expression under a fluorescent microscope.

6. Sort GFP positive cells with FACS machine from BD Bioscience. (Dissociation hESCs into single cells as described in Section 3.1.3). In our experience, hESCs can express DF reporter gene efficiently without changing the expression pattern of differentiation markers (Fig. 3).

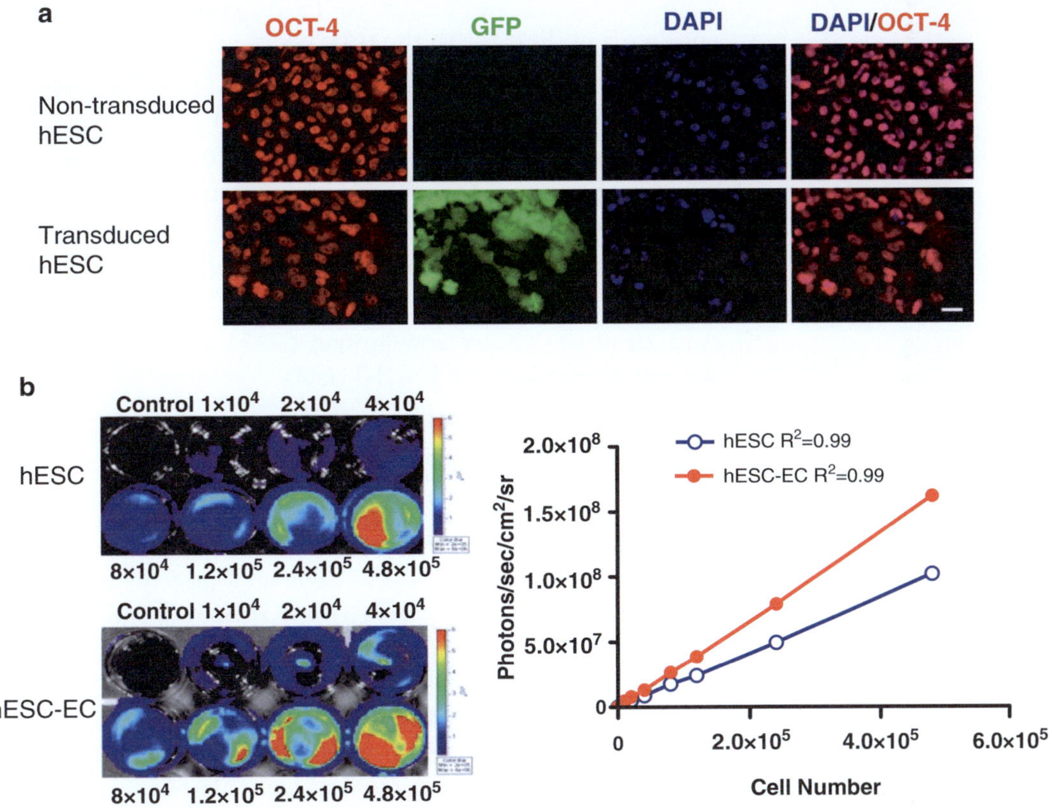

Fig. 3 Stable lentiviral transduction of hESCs with the double-fusion reporter genes. (**a**) Control nontransduced hESCs and transduced hESCs showed similar expression pattern of Oct-4 under fluorescence microscopy. Scale bar = 10 μm. (**b**) Ex vivo imaging analysis of stably transduced hESCs shows increasing bioluminescence signals with cell numbers of hESCs ($r^2 = 0.99$) and with hESC-ECs ($r^2 = 0.99$). Reproduced with permission from ref. (9)

3.3 In Vitro Differentiation of hESC-ECs

3.3.1 hEBs Formation

1. Detach the undifferentiated, infected hESCs as discussed in Section 3.1, and resuspend the cells in differentiation medium containing Isove's modified Dulbecco's medium (IMDM) and 15 % FBS, 1 × BIT, 0.1 mM nonessential amino acids, 2 mM L-glutamine, 450 μM monothioglycerol, 50 U/ml penicillin, 50 μg/ml streptomycin, supplemented with 20 ng/ml bFGF and 50 ng/ml VEGF.

2. Seed the cells in 10 mm ultralow attachment plates, and culture them in 37 °C incubator with 5 % CO_2.

3. Change the culture medium with fresh differentiation medium every other day for 12 days (Note 8).

3.3.2 Flow Cytometry Sorting of hESC-ECs

1. Harvest 12-day-old hEBs and centrifuge the suspension at 200 × g for 5 min at room temperature.

2. Aspirate the supernatant gently, then add 2 ml 0.56 U/ml Liberase Blendzyme IV, and incubate the plates at 37 °C for 20 min with frequent shaking (Note 9).

3. Transfer the cell suspension to 15 ml conical tube, and add another 5 ml EGM-2 to the tube.

4. Let the cell suspension pass through a 40 μm cell strainer, and harvest the obtained single cell suspension.

5. Centrifuge the single cell suspension at 200 × g for 5 min at room temperature. Aspirate the supernatant and rinse the cells with DPBS once.

6. Resuspend the obtained cell pellet in 60 μl DPBS per 1 × 10^6 cells.

7. Add 20 μl PE Mouse anti-human CD31 antibody reagent to the cell suspension per 1 × 10^6 cells.

8. Incubate the mixture on ice in darkness for 30 min.

9. Add 1 ml cold DPBS per 1 × 10^6 cells, and centrifuge the cells at 200 × g for 5 min. Aspirate the supernatant and repeat once.

10. Sort CD31$^+$ cells using FACS machine.

11. Seed the isolated CD31$^+$ hESC-ECs on 4 μg/cm^2 fibronectin-coated dishes. Culture the cells with EGM-2 medium (Note 10).

3.3.3 Subculture of hESC-ECs

1. Take out the culture plates from incubator. Aspirate the supernatant and rinse the cells with PBS twice.

2. For each 10 mm dish, add 2 ml 0.25 % trypsin–EDTA solution to the cells. After cells detaching from the dishes, add 5 ml EGM-2 medium to neutralize trypsin.

3. Collect the cell suspension into a 15 ml conical tube and centrifuge it at 200 × g for 5 min at room temperature.

4. Aspirate the supernatant and resuspend the cell pellet with fresh EGM-2 medium.

5. Transfer the cell suspension into fibronectin pre-coated 10 mm dishes at split ratio 1:3–5.

6. Change medium every other day.

7. For myocardial infarction experimental model, suspend hESC-ECs at the concentration of $1 \times 10^6/20$ μl. (see Section 3.4, step 8)

3.3.4 In Vitro BLI of hESC-ECs

1. Transfer the hESC-ECs suspension (see Section 3.4, step 8) into fibronectin pre-coated 6-well 6 h before BLI.

2. Prepare a 200× D-luciferin stock solution (30 mg/ml) in DPBS and pass through 0.2 μm filter. Use immediately, or aliquot and freeze at −20 °C for future use.

3. Prepare a 150 μg/ml working solution of D-Luciferin in pre-warmed EGM-2 medium.

4. Aspirate media from cultured cells.

5. Add 1× D-luciferin solution to cells just prior to imaging.

6. Imaging for 30 s–1 min (Note 11).

7. Bioluminescence is quantified in units of photos per second per steridian ($p/s/cm^2/sr$) (Fig. 3).

3.4 Myocardial Infarction Model and Cell Transplantation

1. All surgical procedures are performed on 8–10 week SCID beige mice (Charles River, Wilmington, MA, USA). And pay attention to follow the Animal Care and Use Committee guidelines.

2. Anesthetize the mice with 2–3 % inhaled isoflurane for induction.

3. Shave the left side of the ribcage and disinfect using 75 % ethanol.

4. Endotracheal intubation is performed, and mice are ventilated with an animal respirator. Keep the mice in supine position by fastening the legs and upper jaws. And the anesthesia is maintained with 1–2 % inhaled isoflurane.

5. Perform left thoracotomy on the mice. Briefly, cut the skin along the left side of sternum, and detach the subcutaneous tissues along the inferior fringe of pectoralis major muscle. Then cut the pectoralis muscle groups transversely to expose the thoracic cage. After that, enter the fourth intercostal space by blunt dissection and separate the third and fourth ribs using micro-dissecting retractors.

6. Remove the pericardium carefully with scissors and press the thorax slightly to squeeze out the heart.

7. The left anterior descending (LAD) artery is localized 1–2 mm below the junction of pulmonary infundibulum and the left atrial appendage. Ligate the middle of LAD using a 5-0-polypropylene suture from the left border of the pulmonary conus to the right border of the left atrial appendage.

8. For intramyocardial injection, a sterile Hamilton syringe with a 30 G sterile beveled needle is introduced into the base of the heart above the infarction area. 1×10^6 hESC-ECs are injected into three sites near the peri-infarct zone at 20 µl of total volume after LAD ligation. In control group, inject the same volume of PBS solution to the mice.

9. Close the chest cavity in layers using 6-0 polypropylene sutures.

10. Gradually wean the mice from the respirator and remove the endotracheal tube. Monitor the mice until they are fully conscious, and then return them to their cages.

3.5 BLI of Transplanted hESC-ECs In Vivo

1. Prepare a fresh stock solution of D-luciferin at 15 mg/ml in sterile DPBS, and sterile it through 0.2 µm filter.

2. Initiate the Xenogen IVIS 200 system in advance for cooling down the CCD.

3. After disinfecting the abdomen region with 75 % alcohol solution, inject by intraperitoneal route with 150 mg D-luciferin/kg body weight.

4. Place the mice into the induction chamber of Xenogen IVIS 200 system for anesthetization with 2–3 % inhaled isoflurane. Monitor the mice and wait for about 10 min before complete anesthetization and the distribution of injected D-luciferin.

5. 10–15 min later, transfer the mice from the induction chamber to the nose cones in the imaging chamber and close the door (Note 12).

6. Choose the Luminescent option on the control panel. Set the field of view (FOV). Choose the binning, F/Stop, and exposure time (Note 13).

7. Acquire the image by clicking the acquire button. And the bioluminescence signal should be quantified in units of maximum photons per second per cm square per steridian ($photons/s/cm^2/sr$). When the procedure is complete, return the mice to their cage immediately and monitor them until they are fully conscious.

8. The same mice should be imaged for up to 8–10 weeks after operation (Fig. 4).

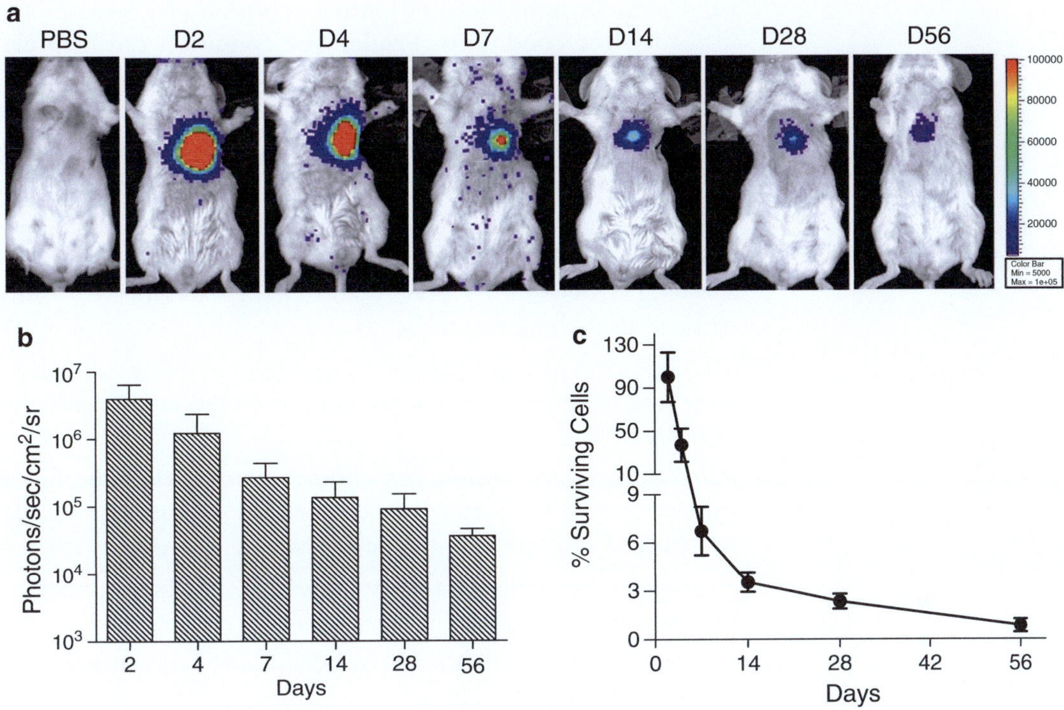

Fig. 4 Molecular imaging of hESC-EC fate after transplantation. (a) A representative animal injected with 1×10^6 hESC-ECs shows significant bioluminescence activity at day 2, which decreases progressively over the following 8 weeks. (b) Detailed quantitative analysis of signals from animals transplanted with hESC-ECs (signal activity is expressed as photons/s/cm^2/sr). (c) Donor cell survival plotted as % signal activity from day 2 to week 8

4   Notes

1. When passaging hESCs, operators have to monitor the cells under a microscope, for long digestion time will damage the cells.

2. Avoid adding penicillin–streptomycin to 293 T medium, which may influence the titer of lentivirus.

3. Plasmids that produce the lentiviral packaging proteins (gag/pol and rev) and the VSV-G envelope protein are commercially available as a ViraPower™ Packaging Mix (Invitrogen).

4. After transfection, 293 T cells may become round and tend to detach from the dish. Thus, operators have to change the medium very gently, or large flakes of 293 T cells will float.

5. Avoid multiple freeze/thaw cycles, which may lower the titer of lentivirus seriously.

6. hESCs can be transduced by lentivirus both on feeder layer and in feeder-free conditions. Feeder-free condition can help elevate the efficiency of transduction for the MEF cells may also uptake a part of plasmid.

7. Try to keep the infection volume as small as possible to insure contact of virus and cell in the allowed incubation time.

8. When change the medium, collect both hEBs and culture medium into 15 ml conical tubes. Wait for several minutes allowing hEBs sink by gravity and change for fresh medium without disturbing the hEBs.

9. The digestion process is harsh. To get high-viability cells, shake plate gently during the process and determine your best digestion time.

10. After this step, operators may choose to verify the function of the sorted CD31$^+$ cells by low-density lipoprotein (LDL) uptake and Matrigel angiogenesis assay.

11. Incubating the cells for a short time at 37 °C before imaging can increase the signal.

12. Luciferin kinetic study should be performed for each mouse to determine peak signal time.

13. Imaging time usually takes 30s–10 min. Operators may choose "Auto" for exposure time for the first time of imaging.

Acknowledgments

This work was partially supported by grants from the National Basic Research Program of China (2011CB964903), National Natural Science Foundation of China (31071308), Tianjin Natural Science Foundation (12JCZDJC24900), NCET of State Education Ministry (NCET-12-0282) and Fundamental Research Funds for the Central Universities (65121018).

References

1. Li Z, Han Z, Wu JC (2009) Transplantation of human embryonic stem cell-derived endothelial cells for vascular diseases. J Cell Biochem 106(2):194–199. doi:10.1002/jcb.22003

2. Wollert KC, Drexler H (2005) Clinical applications of stem cells for the heart. Circ Res 96 (2):151–163. doi:10.1161/01. RES.0000155333.69009.63

3. de Muinck ED, Thompson C, Simons M (2006) Progress and prospects: cell based regenerative therapy for cardiovascular disease. Gene Ther 13 (8):659–671. doi:10.1038/sj.gt.3302680

4. Li Z, Wilson KD, Smith B, Kraft DL, Jia F, Huang M, Xie X, Robbins RC, Gambhir SS, Weissman IL, Wu JC (2009) Functional and transcriptional characterization of human embryonic stem cell-derived endothelial cells for treatment of myocardial infarction. PLoS One 4(12):e8443. doi:10.1371/journal. pone.0008443

5. Moon SH, Kim JS, Park SJ, Lee HJ, Do JT, Chung HM (2011) A system for treating ischemic disease using human embryonic stem cell-derived endothelial cells without direct

incorporation. Biomaterials 32(27):6445–6455. doi:10.1016/j.biomaterials.2011.05.026

6. Yu J, Huang NF, Wilson KD, Velotta JB, Huang M, Li Z, Lee A, Robbins RC, Cooke JP, Wu JC (2009) nAChRs mediate human embryonic stem cell-derived endothelial cells: proliferation, apoptosis, and angiogenesis. PLoS One 4(9): e7040. doi:10.1371/journal.pone.0007040

7. Wang L, Su W, Liu Z, Zhou M, Chen S, Chen Y, Lu D, Liu Y, Fan Y, Zheng Y, Han Z, Kong D, Wu JC, Xiang R, Li Z (2012) CD44 antibody-targeted liposomal nanoparticles for molecular imaging and therapy of hepatocellular carcinoma. Biomaterials 33(20): 5107–5114. doi:10.1016/j.biomaterials. 2012.03.067

8. Su W, Zhou M, Zheng Y, Fan Y, Wang L, Han Z, Kong D, Zhao RC, Wu JC, Xiang R, Li Z (2011) Bioluminescence reporter gene imaging characterize human embryonic stem-cell-derived teratoma formation. J Cell Biochem 112(3):840–848. doi:10.1002/jcb.22982

9. Li Z, Suzuki Y, Huang M, Cao F, Xie X, Connolly AJ, Yang PC, Wu JC (2008) Comparison of reporter gene and iron particle labeling for tracking fate of human embryonic stem cells and differentiated endothelial cells in living subjects. Stem Cells 26(4):864–873. doi:10.1634/stemcells.2007-0843

Methods Molecular Biology (2013) 1052: 217–218
DOI 10.1007/7651_2013
© Springer Science+Business Media New York 2013

INDEX